teaching secondary
SCIENCE USING ICT

Editors: David Sang
         Roger Frost

HODDER MURRAY

Titles in this series:

Teaching Secondary Biology	0 7195 7637 7
Teaching Secondary Chemistry	0 7195 7638 5
Teaching Secondary Physics	0 7195 7636 9
Teaching Secondary Scientific Enquiry	0 7195 8618 6

Papers used in this book are natural, renewable and recyclable products. They are made from wood grown in sustainable forests. The logging and manufacturing processes conform to the environmental regulations of the country of origin.

Orders: please contact Bookpoint Ltd, 130 Milton Park, Abingdon, Oxon OX14 4SB. Telephone: (44) 01235 827720. Fax: (44) 01235 400454. Lines are open from 9.00–6.00, Monday to Saturday, with a 24-hour message answering service. Visit our webste at www.hodderheadline.co.uk.

First published in 2005 by
Hodder Murray, an imprint of Hodder Education,
a member of the Hodder Headline Group
338 Euston Road
London NW1 3BH

Impression number   10 9 8 7 6 5 4 3 2 1
Year   2010 2009 2008 2007 2006 2005

Layouts by Pantek Arts Ltd, Maidstone, Kent

Typeset in 12/13pt Galliard by Pantek Arts Ltd, Maidstone
Printed and bound in Malta

A catalogue entry for this title is available from the British Library

ISBN-10 0 7195 8071 4
ISBN-13 978 0 719 58071 3

Contents

Contributors and editors

Darren Forbes has been teaching secondary science for 10 years and is now Head of Physics at Saints Peter and Paul RC High School in Widnes. He has developed a range of IT resources for science for a number of projects.

Roger Frost is a writer, teacher trainer and software developer. A long-time specialist in ICT and science teaching, he reviews products and creates learning materials for textbook and magazine publishers. Now based in Cambridge, he worked for many years in clinical biochemistry, as an advisory teacher for science, and taught science in East London. His work takes him to conferences across Europe, USA and the Arab states, and he has been developing science software in India for Malaysian high schools. His website (www.rogerfrost.com) is aimed at those seeking advice on IT resources.

Ian Lawrence taught physics and science for nearly 20 years before becoming increasingly involved in curriculum development work, for example leading the electronic developments within the *Advancing Physics* course. He has worked with several modelling tools over the years. He is currently training teachers in Birmingham and co-leading a project to help non-specialists enjoy their teaching of physics.

Rob Musker is an ICT teacher adviser for content development in Lancashire. He has previously been a Head of Science and an ICT Project Manager. He is an author of textbooks on science and ICT and was part of the writing team for *ICT Activities in Science* and *Eureka!* (Heinemann). He has carried out several research projects on the use of ICT in teaching and learning, and contributed his research findings to conferences around the country.

Peter Myers is a practising teacher in Poole, Dorset, where he has taught for 10 years. He is currently Course Manager, Gifted and Talented Coordinator and ICT Network Manager at Poole High School. He teaches all aspects of science to KS4, and Chemistry and Vocational A level Science at KS5. He is an independent evaluator for the *Learn Premium E-valuate* project, evaluating software in the classroom for the Curriculum Online scheme. He has also been an online tutor for the Science Consortium NOF ICT training programme.

Dave Pickersgill is the Digital CoVE (Centre of Vocational Excellence) Development Manager in Sheffield College. He has more than 20 years of experience in teaching and managing in schools and further education, mostly teaching Physics at A level. He has a regular Websearch column in *School Science Review* and also manages the Weblinks part of the Sheffield College website (www.sheffcol.ac.uk/links), which was the winner of a Becta/Guardian Schools/Colleges Website Award. He was recently an 'expert' on the Becta, ICT in Science, *Ask an Expert* website. He is also a moderator for the Becta Teacher Resource Exchange (http://tre.ngfl.gov.uk).

Laurence Rogers is Senior Lecturer in Education at the University of Leicester, having previously taught physics and science in comprehensive schools in Leicestershire. A frequent presenter of papers and seminars at international conferences in Europe, America and Asia, he has published many research papers investigating the use of ICT in science teaching and is joint author with Leonard Newton of *Teaching Science with ICT* (Continuum Books). He is also author of the *Insight* family of Datalogging and Control technology software (Logotron). His experience of datalogging dates back to the era of the BBC Microcomputer when he developed the *Science Toolkit* range of sensors, interfaces and software.

David Sang worked as a research physicist for six years before becoming a teacher. He then taught physics and science for 13 years, in a comprehensive school, a sixth-form college and a polytechnic. He now works on a variety of curriculum development projects, as well as writing textbooks and multimedia teaching materials for students aged 11 to 18. He is physics consultant to PLATO Learning.

Tony Sherborne is a curriculum developer at the Centre for Science Education, Sheffield Hallam University. He has developed multimedia and contemporary science materials for major projects including the *Pupil Researcher Initiative* and *Science Year*. He also directs *Science UPD8* for the ASE.

Acknowledgements

Thanks are due to the following for permission to reproduce screen shots from their software:

CCLRC, Macromedia, Phoenix, PLATO Learning, Sunflower Learning, WorldMaker.

All the websites referred to in this book were active at the time of publication. Although every effort has been made to ensure that website addresses are correct at time of going to press, Hodder Murray cannot be held responsible for the content of any website mentioned in this book. It is sometimes possible to find a relocated web page by typing in the address of the home page for a website in the URL window of your browser.

The cover photo is reproduced courtesy of David Gifford/ Science Photo Library.

Photos in Chapter 2 are by the author, Laurence Rogers.

Every effort has been made to trace all copyright holders, but if any have been inadvertently overlooked the Publishers will be pleased to make the necessary arrangements at the first opportunity.

Introduction

David Sang

This book is the fifth in the ASE/John Murray Science Practice series, accompanying *Teaching Secondary Physics, Chemistry, Biology* and *Scientific Enquiry*. Those earlier books have indicated opportunities for the use of Information and Communication Technologies in science teaching; this book takes a closer look at the various strands of ICT and considers the rationale for using them.

There is no doubt that, in the world of education, technologies come and go. (The language labs of the eighties have all but disappeared today.) However, there is no doubt that ICT is here to stay. Teachers (and some students) carry laptops on which they can gather all manner of electronic resources. Schools have intranets, and science departments have home pages. Data projectors and interactive whiteboards have become commonplace.

Ten years ago, it would have been difficult to write this book. The hardware and software were relatively undeveloped, and the effort required to include any aspect of ICT in one's teaching was rarely justified by the educational outcomes. ICT was the realm of the enthusiast; there were always those teachers who enjoyed playing with electronic gear or writing their own software. (Other teachers may have felt guilty that they weren't doing the same.) Today, we live in a plug-and-play, off-the-shelf world in which the relatively user-friendly equipment and progams make it much easier to think about the educational objectives of our use of ICT, rather than worrying about whether a lesson will grind to a halt as batteries go flat or a computer screen freezes.

Of course, we don't simply advocate the use of ICT in science teaching for its own sake. ICT can transform science teaching because it allows you, the teacher, to do things which can't be done as well in other ways. Static diagrams can come alive, large amounts of data can be gathered, numbers can be crunched, ideas can be represented on a screen, the Internet can be explored. Students can have greater independence in their work and their thinking. In addition, ICT is important because our students are going on to live and work in a world where these new technologies are increasingly important, universal even. Our future scientists and engineers will inevitably find that ICT is a vital part of their working world.

This book contains chapters on the different ICT possibilities in the classroom, written by an experienced team of authors. All have taught using ICT and, variously, have been involved in developing new hardware and software over the last few years, in reviewing new materials, and in helping other teachers to extend their ICT competence.

Our aim has been to encourage you (either individually, or as a science department collectively) to start to incorporate further aspects of ICT into your teaching. While describing straightforward, practical approaches, we have tried to highlight the rationale for using the particular technologies. On the accompanying CD-ROM, we have included many of the software items referred to, as well as provided hotlinks to websites. We hope this will enhance your reading of the book.

On the CD-ROM you will find:
♦ **Chapter 1** Hotlinks related to the ICT toolkit for teachers.
♦ **Chapter 2** A demonstration version of *Datalogging Insight* (functional for 30 days). It can be installed and used to open the files in the 'Sample Files' folder, which contain data for all the experiments featured in this chapter.
♦ **Chapter 3** Hotlinks relevant to using multimedia, as well as three pieces of multimedia: the interactive games *Planet 10* and *Bloodbug* (both from PlanetScience), and *The Heart*, a piece of interactive multimedia developed by PLATO Learning.
♦ **Chapter 4** Files from the *SeeingScience with CCLRC* project which will allow pupils to carry out the PowerPoint activity described on pages 75–79.
♦ **Chapter 5** Instructions of how to create the Excel spreadsheet files described in this chapter, and the final versions of each of the spreadsheets.
♦ **Chapter 6** Installers for the modelling programs *VnR* and *Modellus*, and hotlinks to websites where updates for these can be downloaded, as well as to an installer for *WorldMaker*.
♦ **Chapter 7** Hotlinks for effective use of the Internet.
♦ **Chapter 8** Two PowerPoint files related to the activity on Dihydrogen Monoxide; a template for creating a presentation and a completed presentation. There are also three Word files; two are for use in role plays, and the third is a template to enable pupils to present their findings on Energy Resources.

1 | *The ICT toolkit for teachers*

Rob Musker

♦ *Introduction*

It is now no surprise to find ICT having an impact on teaching and learning in the classroom. The findings of the *ImpaCT2* survey (Becta 2002) provide real evidence of this. The investment nationally is considerable, and where secondary schools have a good ICT infrastructure, this investment can allow the influence of ICT to be felt fully. How much of the overall investment has found itself supporting the teaching of science is unknown, but it appears that many schools and departments have a good idea of how to use technology to produce real learning gains for their students.

This chapter looks at some of the best equipment and software to enhance students' and teachers' learning experiences. Some of the equipment is now commonplace in many science classrooms, while some is not yet found commonly in every lab. Real examples of the use of the hardware tools and software are given throughout the chapter, so you can get ideas to use in your lab now.

1.1 The tools of the trade

To start with, you need to get the hardware right. This section gives some tips.

Getting interactive – using interactive whiteboards

Few tools will affect classroom practice more than interactive whiteboards (IWBs). Their ease of use has ensured that they are now widely adopted in classrooms throughout the country. An IWB system comprises the board itself, a digital projector and a computer. The computer can then be controlled by touching the image on the board either directly or by using a special pen. There are also tools such as Mimio® and eBeam®, which make normal whiteboards interactive.

The main factors that will affect the use of the IWB are:

- the training of teachers in the use of the boards in their subject, and making sure they have good access to them to build up their confidence

- investment in suitable resources for use with the board
- systems that allow teachers to share their ideas and resources on using IWBs
- positioning of the IWB and the digital projector so that they are at the appropriate height for their users, and not too near strong light sources or obstructions
- support of its use with suitable technical help.

Becta (*What the research says about interactive whiteboards*, 2003)

The versatile software that accompanies many interactive whiteboards has many features to support your lessons. The interactivity of the boards can provide more opportunities for classroom discussion, collaboration and student participation. All these can be promoted to a greater extent when a voting system is used. Whiteboards can also provide a very visual, stimulating way to encourage student learning, with the ability to highlight key elements during animations and videos and to save key frames to use later. Perhaps more important is the function that allows teachers to save screens during a lesson and view them at a later date. The screens can even be saved in HTML and viewed via the Internet, thus providing an effective method of producing online resources to support home study.

Digital cameras – picturing the effect

As they say, a picture paints a thousand words. A digital camera can help the science teacher show students how to set up their lab equipment, and can provide an easy way to produce an image for labelling, for example a microscope. There is a wide range of digital cameras to choose from, suiting all budgets. The main criteria for choosing a camera are pixel resolution, cost and ease of use. The ability to record onto floppy disk or CD and then take the pictures away to view on almost any computer makes it possible to share a camera across a whole class.

Digital cameras often come with their own software for manipulating images but some people may want to invest in a more specialised graphics program such as Adobe Photoshop or Paintshop Pro. Once produced, photographs can be incorporated into many different types of software such as Microsoft Office. Digital photography may also help some departments to support science teachers teaching outside their main subject. A database of photographs of equipment and their set-ups can be produced. This allows non-specialists and technicians to see at a glance what equipment is needed for experiments and how to set it up safely and correctly.

Getting closer – using digital microscopes

The most common digital microscope found in the science classroom is the Intel QX3, but there are other options available. The MOTIC DigiScope may give you better picture quality and resolution. All departments should have access to a digital microscope, as promoted by the ASE's *Science Year* initiative. These microscopes not only take pictures but are also able to take videos and time-lapse photographs. Students can photograph or video biological specimens found in leaf litter or look at the structure of rocks, for example. Photographs can be imported into other documents or software packages and labelled.

Digital microscopes can be used to produce close-up photographs of commonly used equipment or chemicals, for example salt crystals or biological samples. These make a ready supply of 'Guess the mystery object' samples that can be used as starter or plenary activities to promote discussion about the scientific concept being studied.

Lights, camera, action and fun – digital video in the classroom

Digital video (often shortened to DV) is video that is recorded in a digital format so that you can store and edit it on a computer. There are two main formats: Mini DV and Digital8®. Mini DV is the most common and uses smaller cassettes while Digital8® uses larger 8 mm cassettes. Generally, the equipment you need to film and edit DV is a DV camera, a computer with a firewire port (also termed as IEEE1394 or iLink) and editing software. More information on digital video in teaching and learning can be obtained from Becta (*Using Digital Video in Teaching and Learning*, 2002).

Note that there are also many digital cameras, for example Sony Mavica®, that come with some facility to produce short digital video clips. The clips can then be transferred to a computer and viewed. Generally these cameras produce AVI files, which may limit the choice of DV editing software.

There are a huge number of different types of camera and choices of editing software. Some editing software comes free with your computer or operating system, such as iMovie (with Apple Macs) and Movie Maker (with Windows ME or XP). There are also many other editing packages, some of which are more suitable for the more expert user, e.g. Adobe Premiere, while others are more suitable for all users, e.g. Pinnacle *Studio*.

DV can be used in a most creative manner by teachers and students. Some ideas for the use of DV in the science classroom are outlined in Table 1.1.

Table 1.1 *Teaching science using DV.*

Activity	Explanation
Teacher or student presentations	Students can produce role plays (see below), science songs, virtual dissections, virtual experiments or presentations on science topics. For the teacher, DV is particularly useful in reviewing topics, in teaching difficult scientific concepts or producing a visual copy of those experiments that never seem to work. Another idea is an interactive periodic table: when the element is clicked on, a student tells the viewer about the element.
The production of plays or role plays	Plays are an excellent chance for cross-curricular work. They work particularly well for emotive topics that include aspects of citizenship – such as pollution, organ transplants or genetically modified organisms. Videos of the common role plays such as electrical circuits or particles can be replayed to the students to aid revision.
Observing student teachers	It is so much easier to show somebody what happened and, more importantly, played through a computer it provides an instant review of what happened. The student teachers could make a virtual portfolio to show their progress during the year.
Observing and analysing discussion and thinking skill lessons	This enables the teacher to analyse lessons and share good examples of thinking with the students and other teachers, in order to develop strategies to improve scientific thinking.
Other presentations	Videos of the department to show to other schools during briefing sessions or to parents during Open Days can produce an excellent impression – you choose what they see!

Case study: an e-soap opera

As part of a summer school for gifted and talented students in 2003 at Fulwood High School and Arts College, students produced an e-soap called 'Future Shock' which dramatised environmental problems such as the hole in the ozone layer. The e-soap, a film in four episodes, was broadcast on the school intranet over a four-week period. The students were from Years 6 and 7 and this was the first time they had used DV. They filmed and edited the drama as well as composing the soundtrack and writing the script. This example shows how creative and accessible DV is, even to the complete beginner. More information on the e-soap project can be found on the Lancashire Grid for Learning website (www.lancsngfl.ac.uk).

The downside

Needless to say, the main problems that arise with DV involve the lack of sufficient equipment for filming or editing to provide resources for even small group work. However, the resourceful teacher can accommodate even this problem by using a circus of activities or by rotating use throughout the year.

DV users should also take into consideration the child protection issues regarding the storage of any films with student participation. See *Using Digital Video in Teaching and Learning* (Becta 2002) and the website *Superhighway Safety – Safe use of the Internet* for guidance.

Using sound

It is easy to record sound and use it creatively in science lessons. A cheap microphone and a computer in conjunction with sound recording software such as Microsoft's Sound Recorder, which is free with Microsoft operating systems, makes it easy to use and incorporate sound into other presentations or documents. Some science departments have a 'bookable' set of microphones.

Some ideas for activities using sound are given in Table 1.2.

Table 1.2 *Using sound in science activities.*

Activity	Implementation
Labelling activities	Students drag sound files to the correct part of the diagram or they label their own diagrams using sound files. This can be done using many different types of software, e.g. Microsoft Word or PowerPoint, MatchWare Mediator or Macromedia Flash.
Scientific experiments	Students record part or all of their scientific experiments using sound files. These can then be incorporated into documents, e.g. Microsoft Word documents.
Radio (audio) plays/ presentations	Students can produce plays, perhaps on emotive issues, or record plays that exist already for discussion purposes (e.g. 'Give me Sunshine' from ASE's *Science Year* CD3. Sound files can be placed into authoring software (e.g. Microsoft PowerPoint, MatchWare Mediator or Macromedia Flash) with accompanying graphics or animations.

Making it live – video conferencing and webcams

There are many different systems you could use to link up your students with scientific experts, teachers or other students in this country and across the world. You will need a suitable camera, software, and willing partners who have compatible equipment for video conferencing. Many people use Microsoft's Netmeeting to converse via the Web as it is very accessible and a free piece of software. It allows you to share your computer with somebody else, who will be able to look at documents, sounds or pictures on your computer.

Webcams and suitable software allow you to put live pictures on the Internet for other people to view, anywhere in the world with Internet access. This allows you to show and share live experiments, either as a photograph that refreshes after a set time period or as streaming video. Pictures with many less expensive webcams are adequate but if you need better photographs or video you will need to get a more sophisticated webcam. There are webcams that you can control via the Internet, i.e. you can pan, tilt or zoom the camera.

Lancashire Grid for Learning has broadcast simple experiments showing the growth of plants and crystals. Pictures were taken at regular intervals which allowed time-lapse films to be produced to show the entire experiment in about 60 seconds. Another series of experiments looked at different levels of fertiliser and light. These are common experiments for secondary students following courses with an applied or vocational character.

Antcast at the Natural History Museum shows the day-to-day antics of leaf-cutter ants, and there are numerous websites showing live pictures from reefs or marine life centres.

A portable solution

There are always many decisions to be made where ICT is concerned and one of the main ones is choosing whether to have a portable computer or a desktop model. Both options have advantages and disadvantages. Portable computers such as laptops appear to have a shorter 'life expectancy' but provide a very flexible ICT solution, especially when used in conjunction with a wireless network. Previously the deciding factor in this choice was an economic one but now this is not a major issue. There are also other portable solutions that you can choose.

There are many niche products that can differ in functionality and size. Amongst the smallest of these devices are those that are now collectively termed Personal Digital Assistants (PDAs). PDAs, or Pocket PCs, are not new additions to the lab, having been seen as a tool for datalogging for some years. Data Harvest's Flash Logger provides a very cost-effective portable datalogging solution but a PDA can do much more. As this technology evolves and the specification of the machines gets better, PDAs will have a great impact on teaching and learning.

Many PDAs have the capability to run Microsoft PowerPoint, Macromedia Flash animations and other useful document types, meaning that with the aid of a projector you can carry your lessons around with you. The devices are also useful for registering classes. They have a built-in microphone and do all the personal organisation for your work while fitting neatly in your pocket. Many of the hybrid PDA/phones also take digital pictures, and with wireless capability in many devices the flow of information to and from the school network or the Internet is possible.

There are other options that are worth consideration, such as those devices that are smaller than laptops but have very similar or equal functionality, e.g. Tablet PCs. These act like portable clipboards that are light and easy to carry around. They will generally do everything a PC can do but are controlled using a stylus on the screen. A wireless Tablet PC will allow the user to save work directly to a network, and when used with a projector can provide a real alternative to an IWB. Those people who find the use of the stylus difficult can invest in a keyboard, and external drives can be bought for floppy disks, CDs and DVDs. Those Tablet PCs with a Compact Flash Slot can also be used with Data Harvest's Flash Logger, providing a flexible method for datalogging.

Bringing IT to life – data projectors

A data projector can bring to life even the simplest of ICT lessons such as using a CD-ROM or showing a digital photograph. Most of the applications of ICT discussed so far either benefit greatly from (or, in the case of interactive whiteboards, rely entirely on) the use of digital projectors. Interestingly, digital projectors in conjunction with a wireless mouse and keyboard can provide a cost-effective way of making your whiteboard interactive from anywhere in your classroom. The main factors affecting the choice of data projector include initial cost, running costs, weight, light output or brightness, resolution, and whether it has extra features. Nowadays you will also have to consider whether it is beneficial to have a wireless projector. With no wires needed to attach it to your computer this makes it even more flexible and much easier to set up in different places.

1.2 Software

In this section we will look at the use of software in the science classroom. First we will give an overview of the types of software that you can buy and use to support your lessons, and then we will look more closely at some generic software packages that teachers and students can use in science lessons.

The choice is yours – an overview

There are now many excellent pieces of software to choose from to support all facets of the science curriculum. There are software packages that give you a complete solution, such as *Absorb Physics* or *Absorb Chemistry* from Crocodile Clips, which cover the whole curriculum for these subjects with a blend of text, animations, images, videos and self-assessment tests. There are also many pieces of software that are made up of simulations which allow students to manipulate the variables to study the relationship between them. There are many good examples, such as Sunflower Learning's *Multimedia Library for Science* or PLATO Learning's *Multimedia Science School*. These provide some excellent examples of how difficult concepts can be made easy once students can visualise what's happening (e.g. dissolving), and it can be used to promote more discussion and thinking about these scientific concepts.

Immersive Education's *People in Science* allows students an easy way to study some of the most important scientific discoveries, thus supporting scientific enquiry in a very creative manner. Focus Education produces several products such as *Scientific Investigations* which support scientific investigational work by allowing students to view the experimental set-up and methods and, in many of them, manipulate the variables to find out what happens in the experiments. Most of the key experiments at Key Stages 3 and 4 are covered in these packages.

There is more about multimedia software in Chapter 3. Datalogging and its software are dealt with in Chapter 2 and online resources in Chapters 7 and 8.

'Jack of all trades' – generic software

Generic or non-subject specific software can be used by teachers to produce resources or by students to produce their own resources or presentations. Everyone will probably be familiar with MS Office and many with Macromedia Flash, but there are other generic pieces of software that are well suited to use by science teachers and students. Useful software packages are Macromedia's *Dreamweaver* and *Director*, *Hot Potatoes* (see page 13), MatchWare products such as *Mediator* and *OpenMind*, and *HyperStudio*®. Here we will give an overview of some of the ways in which some of this software can be used effectively in the science classroom.

Authoring tools

Many of the most effective pieces of software available for the science classroom are those used for authoring work such as presentations or websites. MS PowerPoint is the most well known of the tools, having been a firm classroom favourite with teachers and students alike for several years. The newer versions of the software make it easier to animate objects and contain numerous extra features to enhance presentations. There are other useful products such as *Mediator* from MatchWare or, for the more technically minded, some of Macromedia's products. With *Mediator*, I particularly like the different formats that presentations can be saved in, such as Flash and HTML. Authoring tools can generally be used for the following activities.

Presentations

Students can produce presentations to introduce or review topics, and teachers can use presentations to support the teaching and learning of topics. Students can be aided in the production of their presentations if they use writing frames. In PowerPoint hidden slides (slides not seen when it is viewed) can be added to the template to give extra support for those who might need it.

Starter or plenary activities

Teachers can produce or purchase short presentations to introduce, review or provoke thought or discussion about a topic. Figure 1.1 shows part of an animated cartoon that promotes thought and discussion about weight and mass. This simple activity is made in Macromedia Flash, as this format produces very small file sizes and is suitable for using online. The activity is primarily intended for using in conjunction with an interactive whiteboard but it can also be used by students working alone, as extra programming allows feedback to be given regarding the correct answer.

Figure 1.1
A cartoon to promote discussion, produced using Macromedia Flash.

Animations

Software can be used to produce animations to help visualise scientific concepts such as diffusion, or to study moments. Later versions of PowerPoint allow you to show the path through which you want to move the object, which saves a great amount of time. Animation in older versions relies on the user producing a sequence of screens to animate an object, much like sequences in a cartoon. Many of the animations and simulations commercially available will have been produced using Macromedia Flash. Although it may take you more effort to produce them yourself, the results are often worth the extra time. For a detailed example, see Chapter 4.

Websites

Many of the authoring tools already discussed can be used to make websites. In fact the primary function of some of the software already mentioned is for making websites, e.g. Macromedia's *Dreamweaver* or Microsoft FrontPage, while others such as MatchWare's *Mediator* or Macromedia Flash are very versatile and can also be used to make effective websites. MS Office documents can all be saved in a format that is suitable for viewing on the Internet. MS PowerPoint and Word can even be used with hyperlinks to make websites.

Microsoft Word

For several years, MS Word has been much more than a word processing package. Here are some examples of adventurous ways to make use of its extensive powers.

Scientific reports

Graphics, photographs, sound and even video can easily be inserted into documents. This allows students to produce multimedia documents.

Interactive worksheets

There are several useful features of Word that make it suitable for producing worksheets. Some of the most commonly used features are drop-down form fields, hyperlinks, comment boxes, and the ability to move text and objects around the screen.

Drop-down form fields accessed on the Forms toolbar provide the user with the ability to make worksheets with a list of possible answers for a set of questions or 'copy and complete' paragraphs. An activity on the eye (Figure 1.2) shows how this technique is used for labelling diagrams. Note that the lists will only work if the document is protected using the *protect form* function from the Form toolbar.

Figure 1.2
An activity sheet developed using Word.

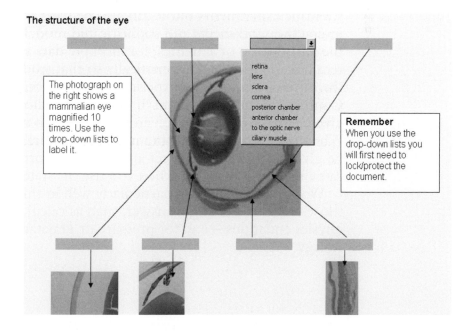

Hyperlinks allow access to other documents on a computer or, if an Internet link is available, to websites. This allows students to link to suitable resources chosen by the teacher to support their work.

Comment boxes can be inserted into documents to provide extra information to support students in their work. They are useful for providing an on-screen glossary of the more difficult scientific words or instructions. Sound as well as text can be used with this feature.

Text or objects such as images can be moved around the screen. This can be achieved by highlighting them and clicking on them with the right mouse button. The highlighted text or object can then be moved anywhere in the document. This is useful for ordering exercises or for 'mix and match' resources.

Other useful features of this software include online collaboration, macros and mail merge, which are all worth knowing about.

Microsoft Excel

MS Excel is a very versatile piece of software. Features such as conditional formatting, 'if' statements, macros and the chart wizard make Excel a valuable tool in making many different types of resources.

Commonly Excel is used to graph data produced by scientific experiments but it can also be used to produce interactive worksheets and sophisticated models. Macros allow the automation of activities, for example data sets can be recalled and graphed automatically so that students can study the differences between experiments at the touch of a button. Conditional formatting and 'if' statements allow the teacher to give feedback to answers given by students in worksheets. When conditional formatting is used, feedback to students can be given as green for correct and red for incorrect, while you can choose the wording when you use 'if' statements.

Worksheets that work particularly well in this format are those relying on a single answer, such as calculations for physics equations – e.g. for pressure or gravitational potential energy.

Hot Potatoes™

This is an excellent software package that can be downloaded from Half-Baked Software and used free of charge by educational establishments (subject to a licensing agreement). The exercises need a browser such as Microsoft Internet Explorer or Netscape® Navigator to work and can be used successfully on- or off-line. *Hot Potatoes*™ is made up of six different types of interactive exercise or application, as outlined in Table 1.3. The exercises are easy and quick to make with the online help providing useful tips on how to produce and customise your work. Feedback is given to students as a percentage score, with every attempt made by the student being taken into consideration. Lessons that involve students in producing their own tests and exercises are also very effective.

Table 1.3 *Using* Hot Potatoes™ *in science lessons.*

Types of Exercise	How to use them
[JQuiz]	You can choose from several different types of quiz such as multi-choice or short-answer questions. The exercises could be used to introduce or review any topic. They work well when used on an interactive whiteboard.
[JCloze] Fill-in-the-blanks exercises	These are fill-in-the-gap exercises that can be used in any lesson to gauge what has been learnt. They are particularly suitable for less able students, especially as the software also allows you to give clues to the answers.
[JMix] Jumbled sentence exercises	These exercises rely on the students rearranging a set of jumbled letters to reveal a word, or jumbled words to make a sentence. These can promote thinking and discussion about topics and are particularly useful at the end of a lesson to review the key words or definitions.
[JCross] Crossword exercises	Crosswords can be used for any topic and are particularly useful for review purposes. They can easily be made for different abilities of student and also help promote scientific literacy.
[JMatch] Matching exercises	There are two versions, a standard and a drag-and-drop version. Students can use them for ordering exercises such as the rock cycle or the metals in the reactivity series. They can also be used to match words with their definitions or link pictures to their name or function, e.g. matching the organs of the digestive system with their graphic or function.
The Masher	This is a flexible tool for making your exercises and other web pages into units of work. The software will link the exercises together via navigation buttons and produce an index page.

◆ *Conclusion*

This chapter has given you an overview of the choices you may have to make regarding ICT in a science department. There are many new products appearing all the time which will influence the way science is taught in the future. The full effects of wireless technology have still not come to fruition, while PDAs or Smartphones with greater functionality are being released and it is only a matter of time before they become commonplace in classrooms and pockets everywhere. What you have to decide is which tools and software are best suited to your teaching and which will enhance the learning experience of your students.

◆ *References*

Becta (2002) *ImpaCT2* survey:
ImpaCT2: Learning at Home and School: Case Studies
ImpaCT2: Pupils' and Teachers' Perceptions of ICT in the Home, School and Community
ImpaCT2: The Impact of Information and Communication Technologies on Pupil Learning and Attainment
    www.becta.org.uk
Becta (2002) *Using Digital Video in Teaching and Learning*
    www.becta.org.uk/corporate/publications/documents/
    Using%20Digital%20Video.pdf
Becta (2003) *What the research says about interactive whiteboards*
    www.becta.org.uk/research/reports/docs/
    wtrs_whiteboards.pdf
Superhighway Safety – Safe use of the Internet website:
    Guidance on using images and digital video on school websites.
    http://safety.ngfl.gov.uk/schools/document.php3?D=d74

◆ *Resources*

Antcast
  www.nhm.ac.uk/museum/creepy/antcastintro.html
Antcam
  www.antcam.com
Half-Baked Software
  www.halfbakedsoftware.com
Hot Potatoes™ website
  www.halfbakedsoftware.com/hot_pot_website.php
Science Year online resources
  www.planet-science.com
Science Year CD3 'Can We Should We?'
  Association for Science Education Booksales
  College Lane, Hatfield, Herts AL10 9AA
  www.ase.org.uk/htm/book_store

2 ICT for measurement: datalogging

Laurence Rogers

◆ Introduction

As a contribution to students' experience of science in secondary school, practical laboratory activity has long had an important role and, in recent years, 'datalogging' has emerged as a distinctive application of ICT to this area. Before considering the nature of datalogging, it is worth giving some thought to the rationale which underpins much of the practical work conducted in schools. There are several strands of argument which support the role of practical work (see Wellington, 1998) and in my view a key factor is that observation and measurement lie at the heart of obtaining the evidence from which scientific thinking can flow. Much of science teaching attempts to develop students' skills of first-hand observation with their primary senses, but it is also the case that many instruments for 'extending' our senses (telescopes, microscopes, meters of many kinds, oscilloscopes, to name a few) have evolved, and science teaching has accordingly devoted attention to developing students' competence in the use of such instruments. Beyond observation, when we consider measurement, instruments of some kind have always been needed. Measurements add precision to observation and pave the way for refining observations, theories, predictions and so on. In the best examples of practical science, the activity seeks to make explicit links between the phenomenon under investigation on the bench and students' understanding of the concepts involved.

Like it or not, technological advance has rendered some older measuring instruments obsolete and this poses a dilemma for teachers having to train students in the use of instruments. It becomes necessary to consider when skill with a particular instrument loses currency. This is the sort of challenge brought about by the new wave of measurement technology known as 'datalogging', and it prompts a number of questions:

◆ What are the novel qualities of datalogging?
◆ What instruments and methods are rendered redundant by datalogging?

♦ Is there a role for the co-existence of old and new instrumentation?
♦ What are the choices to be made?
♦ What are the criteria for judging 'appropriate' use?
♦ What are the risks, potential pitfalls and costs in terms of time and training effort needed for success?

This chapter attempts to help the reader find answers to these questions by a discussion focused on a selection of datalogging activities, chosen to exemplify the range of opportunities useful to secondary school science.

2.1 Datalogging for teachers and students

The term 'datalogging' was coined to describe the process of gathering and recording measurement data, but it is also intimately connected with display and analysis on a computer screen. Thus, when datalogging equipment (sensors, interface, computer and software) is assembled correctly, the data handling process appears to involve four themes: collection, storage, display, analysis. Each has a counterpart in conventional practical work: taking readings, writing them down, drawing a graph, drawing conclusions from the graph. For the latter, teachers have well rehearsed methods of developing students' competence, but how do teachers interact with the processes for datalogging? Despite the apparent dominance of the role of the computer, there is much to occupy the teacher in a pedagogical role:

- **Collection** of the data is usually automatic during the experiment, being controlled by the computer, but the hardware needs to be set up and software parameters for collection chosen at the outset.
- **Storage** of the data is also automatic in the computer memory during the experiment, but data usually needs to be secured for subsequent analysis (saved as a computer file), just as the products of word processing need to be saved as electronic files on a hard drive, floppy disk or CD.
- **Display** of data is usually automatic on the computer screen, but choices need to be made about the format: digits, table, bar chart, YX graph; and about scaling and axes.

- **Analysis** of the data demands the active involvement of students in finding out about the data; this requires awareness of tools, vision and understanding of their purpose, their method of use, and the skill to operate them. Analysis tools need to be employed within a framework of tasks with clear objectives defined by the teacher.
- **Interpretation** of the data follows on from analysis but the computer can have only a minor role in driving this, leaving the main responsibility to the teacher in helping students to understand the significance of information, to associate it with their previous knowledge and to draw sensible conclusions.

Although, superficially, datalogging might appear to displace students' activity by performing tasks traditionally performed by students themselves, in reality there are many compensatory activities that have the potential to enhance students' thinking about and understanding of science. In particular, datalogging can be a valuable facilitator of an investigative approach to practical work, in that it reduces the need for certain low-level gathering and recording skills but at the same time offers a range of analysing methods which support exploratory higher-order thinking. For teachers, there may be a temptation to adopt the role of observer when computers are in action in the classroom, but much research has underlined the importance of teacher interventions in securing learning benefits from computer-based methods (see Rogers and Finlayson, 2003). This always requires clarity of learning objectives and sometimes requires fresh thinking about what those objectives are. The following examples of activities will illustrate learning opportunities for students and action points for teachers.

Informal experiments for gaining a 'feel' for sensors

As a prelude to substantive discussion of some datalogging experiments, we will consider how teaching can address the potential problem of anonymity of measurements with sensors. There is a risk that, since the measuring process for a wide variety of electrical sensors has the same appearance for diverse physical phenomena, the identity of the physical quality being measured becomes ambiguous. In other words, there is a certain 'sameness' about all electrical sensors: they typically have a 'probe' connected to an interface and the product appears as a set of digits or graph line on the computer screen. Lost is the individual personality possessed by conventional sensors such as glass thermometers and analogue voltmeters. With these, as well as the distinctive physical appearance of the instrument, which tends to become associated with a physical phenomenon, the methodology of use is also quite distinctive; the technique of reading a thermometer feels quite different from that of the voltmeter needle located on a graduated dial.

For sensors used in datalogging, how does the teacher introduce a 'feel' for the physical quantity in question? How do you help make the sensory connection between the phenomenon under investigation and the output on the screen, be it a graph or number? How do you help students appreciate what the sensor is doing in the measuring process? One solution is to allow or encourage informal exploratory activity with sensors before conducting the main experiment. Here are some ideas frequently used in a variety of contexts.

Temperature

1. Set up a temperature probe and software to present a 'real-time' graph display. Just holding the temperature probe in the hand or placing it against the side of the neck produces a prompt response on the screen. Rubbing the probe across the palm of the hand several times can generate temperatures well in excess of normal body temperature. When two probes are available, two students can perform this simultaneously and have a 'temperature race', with the results being simultaneously compared on the graph.
2. Discuss with the class what would happen to the reading on the screen when a temperature probe is inserted into a bottle containing a volatile liquid such as methylated spirit. How does the result compare with the prediction? Insert two probes into the liquid and speculate what will happen when they are removed into the air again. Compare the result with the prediction. Speculate about the effect of shaking one of the wet probes around in the air. Again, observe the graph to compare the result with the prediction (see Figure 2.1, overleaf).

These simple demonstration experiments only take a few minutes but they can be a rich source of speculation, questioning and discussion. "What limits the fall in temperature? What affects the rate of fall? How does the average temperature for each probe differ?" and so on. The cycle of predict–test–evaluate can be repeated at will.

Figure 2.1
*A real-time
display of
temperature
readings.*

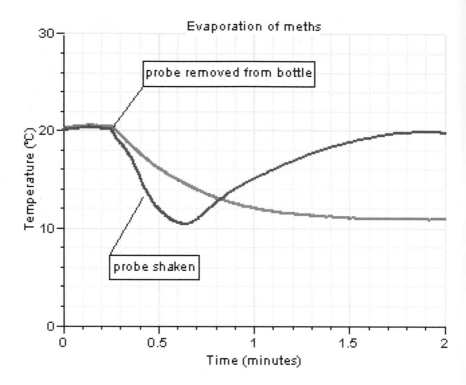

Figure 2.1
A real-time display of temperature readings.

The experiments also provide useful illustrations of these novel features of 'real-time' datalogging:

- The 'results' are immediately observable on the screen. This is the essence of real-time experiments where the results are presented simultaneously with the experimental activity.
- In subsuming the normally laborious business of manually recording data and plotting a graph with pencil and paper, the computer promotes the graph as a starting point for thinking and discussion rather than being the end product, as in a conventional measurement activity.
- The rapid reporting on the graph facilitates an interactive approach to the experiment. "Let's see what happens if....". This is in the best spirit of an investigative, inquiry-based approach.
- It is possible to associate the phenomenon immediately with a feature of the graph. Students quickly learn about the 'language' of graphs for representing and describing change and for comparing data.

Light

1. Set up a real-time graph display with a light sensor connected to the interface. Point the light sensor in turn towards the ceiling, wall, window and floor. Observe the different readings. Then repeat the exercise in a different sequence but with students' eyes closed. When they are allowed to open their eyes, students should be able to guess the sequence by interpreting the new set of readings implicit in the graph.

2. Connect a light sensor to a datalogger set up to work remotely, i.e. disconnected from the computer. Switch the laboratory lights on, hold the sensor vertically, pointing towards the ceiling, start logging and walk at a steady speed in a straight line across the width or down the length of the room. Speculate with the class what shape of graph might be expected. Download the data to the computer and observe the graph (Figure 2.2). Discuss the results and compare with the prediction.

Figure 2.2
A display of light level readings downloaded from a datalogger.

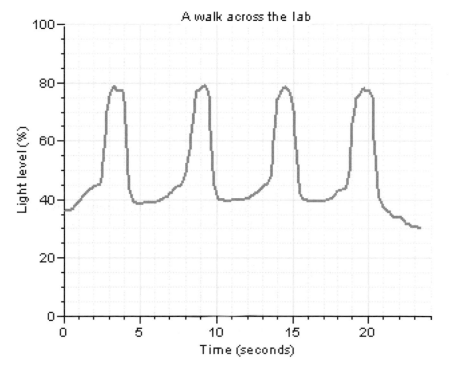

In addition to the features noted for the temperature experiments (page 20), these experiments illustrate further novel aspects of datalogging:

- The graph can be used for simple analysis tasks. Graphs are valuable tools for representing, storing and analysing information.
- Remote logging allows data to be collected in new contexts where a computer is not necessarily present.
- After a remote logging activity, the datalogger must be connected to the computer and the data downloaded for viewing on the screen. Most datalogger systems can store data from several different experiments.

Sound

1. Set up a real-time graph display with a sound level sensor connected to the interface. The graph needs to have a time axis spanning several seconds. Clap your hands twice in fairly quick succession. The time interval between the claps is observable from the graph.
2. Fix a sound level sensor near an electric kettle containing some cold water. Set the datalogger running with a real-time display over a few minutes, switch on the kettle and observe the graph line gradually building up as the kettle comes to the boil. It is necessary to have a quiet class during the experiment and all eyes on the screen! After the experiment, discussion can focus on how the noise made by the kettle moved through distinctive phases which show up on the graph (Figure 2.3).

These experiments illustrate further novel features of datalogging:

- Datalogging is useful for recording changes in much more detail than occurs when we rely upon our primary senses. In this case, the crescendo of noise made by the kettle produces a characteristic shape of graph line.
- The graphing software usually provides a tool for obtaining time interval measurements with great accuracy.

Figure 2.3
*A real-time
display of sound
level and
temperature
readings.*

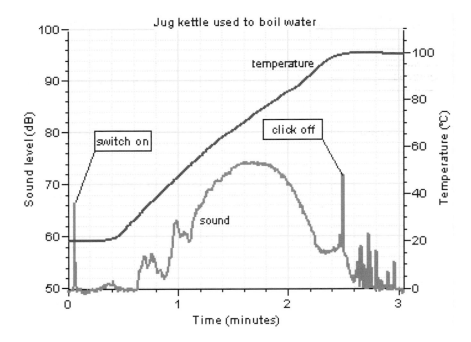

Humidity

1. Set up a real-time graph display with a humidity sensor connected to the interface. Slowly breathe in and out with the sensor held in front of the mouth. Discuss the reasons for the distinctive up-and-down movement indicated on the graph. This provides an opportunity to apply students' knowledge of the process of respiration and the difference between inhaled and exhaled breath in terms of the water vapour content.

2. At the beginning of a lesson, connect a humidity sensor to a datalogger set up to work remotely; it does not need to be connected to a computer. Close all the windows which might normally be open. Halfway through a lesson deliberately open the doors and windows to ventilate the room. Towards the end of the lesson connect the datalogger to a computer and download the data. Observe the graph (Figure 2.4) and discuss what may be learned from its shape.

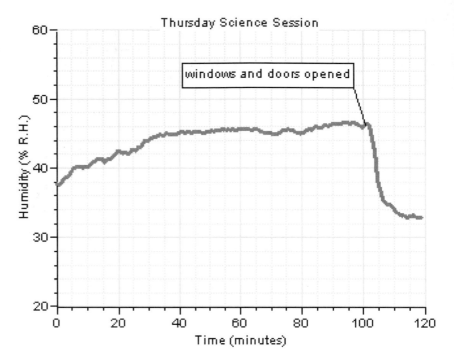

Figure 2.4
A display of humidity readings downloaded from a datalogger.

These experiments illustrate further novel features of datalogging:

• Some sensors make 'visible' what is often invisible to our normal primary senses.
• In remote collection mode, dataloggers are useful for environmental measurements spread over longer periods of time than are normally convenient in timetabled lessons. Some dataloggers have temperature, light, pressure and humidity sensors built in.

Use of a light gate

1. Set up a real-time bar chart display with a light gate sensor connected to the interface. A light gate is different from the previously mentioned sensors in that it sends the computer a digital rather than variable signal; it is either 'on' when the light beam is uninterrupted or 'off' when the beam is blocked by an opaque object. Its properties are simply demonstrated by passing a hand of outstretched fingers perpendicularly through the beam. For each finger a bar appears on the chart on the screen, effectively counting how many fingers passed through the beam (see Figure 2.5).

Figure 2.5
A bar chart display of light gate 'off' time. Four fingers passed across a light gate four times.

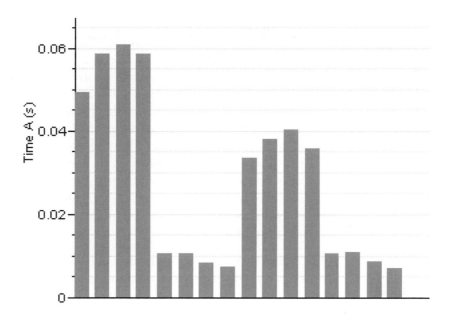

Furthermore, the transit times for each finger are recorded with great precision on the screen. Measurements of this type can be used to prompt discussion of how scientists are able to measure a variety of rapid events and the motion of fast-moving objects.

2. When two light gates are available, two students can experiment to find their reaction times to a visual stimulus. Arrange a screen between the two students so that neither can see the other's hands but both can see the computer screen. One student passes a hand through the beam of one light gate, and the computer screen simultaneously shows that timing has started. The second student, on seeing this, attempts to interrupt the beam of the second light gate as soon as possible. This action stops the timing, and the clock reading on the screen shows the time for the second student to react. Repetition of the experiment can demonstrate the effect of practice on the students' reaction times.

These experiments illustrate how a datalogging system can count and time events with great precision, a property that forms the basis of a large range of measurements of motion. The computer program can be set to perform 'instant' calculations on incoming data to yield measurements of speed, acceleration, momentum and kinetic energy.

pH measurements

Set up a real-time bar chart display with a pH probe connected to the interface. The program needs a facility for recording and plotting single readings from the probe. Organise a series of plastic cups, each containing a different common drink such as milk, orange juice, lemonade, cola. Place the probe in each drink in turn, prompting the program to record pH readings and present them on the bar chart. Take care to rinse the probe in de-ionised water after each test.

In this application the datalogging system is mainly being used as a pH meter. However, it does facilitate a whole class display and the bar chart allows prompt comparison of the series of recorded values.

All of these informal experiments can help build students' confidence in the function of sensors as reporters of real physical phenomena, and they are recommended as introductions to datalogging and the use of common sensors. They also serve to illustrate how simple ideas can be used to achieve a high profile for scientific discussion rather than undue distraction with technical adventures. There is an ever-attendant risk of students and teachers, in engaging with the obvious need to make things work, becoming preoccupied with the operational needs of the technology at the expense of thinking about the scientific ideas. For the novice user, it is wise to begin with modest ambitions with the technology itself, focusing on simple uses so that the science is to the fore. Fortunately the design of modern equipment and software has generally sought to minimise the technical overhead, but it is important to strive for an appropriate balance between the science and operational factors.

New versus old

The novel features of datalogging noted in the previous examples serve to illustrate aspects of the 'added value' that datalogging can offer beyond measurement with equivalent conventional instruments. As with any teaching aid, it is important to be clear about the advantages and disadvantages of the tool and to match these against teaching objectives. Sometimes the benefits of datalogging and conventional methods rival each other, such that the former is a mere alternative to the latter. One-off measurements usually fall in this category, where the business of organising a sensor and datalogger is not merited when a simple instrument would suffice. For example, the humble glass thermometer is the sensible solution when only one or two temperature readings are needed. However, if you wish to observe small *changes* in temperature, the sensitivity of a temperature sensor offering readings to a precision of 0.1 degree offers a distinct advantage; or if you wish to observe a *rate* of change, a sequence of readings in close succession is more easily recorded with a datalogger.

There are many instances when the datalogging method is completely novel and cannot be replicated with conventional instruments. The real-time interactive graph display in Figure 2.1 is a clearly recognisable example of this, with the graph being updated simultaneously with the experimental activity. The examples later in this chapter will illustrate further ways in which datalogging offers distinct advantages. It is hoped that these examples will help readers to seek out 'added value' in a wider range of applications. Without this 'added value' the justification of datalogging methodology is difficult to sustain. It is important to identify and exploit this value where it exists (see page 28).

There will also be instances when datalogging is the wrong solution for a measuring task; for example, rulers, weighing balances and thermometers reign supreme in many everyday laboratory tasks. It is appropriate to recognise that sometimes traditional instruments are not only more convenient but their first-hand nature confers understanding, as in the case of a ticker-timer for measuring speed with ticker-tape; both of the essential elements of speed calculation, distance and time, can be extracted directly from the markings on the tape.

Reflecting again on the novel features of datalogging noted in the experiments above, we can identify a mixture of two types: system properties and learning benefits. For example, 'automatic graph plotting' and 'remote logging' simply describe properties of the system which may be an advantage to the learner, but their use does not guarantee successful learning. In contrast, 'facilitates an interactive approach' and 'promotes the graph as a starting point for thinking and discussion' can be construed as benefits to learning. In both cases, the achievement of successful learning depends upon the skill of the teacher in training students to use the tools productively, designing the task with clear objectives in mind and managing the activity in the lesson. We shall return to this theme through the examples that follow.

Overall, an important role of the teacher in planning datalogging activities is to invoke continually the principle of 'fitness for purpose'. Does the application promise to deliver the learning objectives? How appropriate is the datalogging system in terms of its sensitivity, accuracy, robustness, cost, complexity and convenience? Does the task exploit the 'added value' of the method? What skills are needed for the activity? How can the task be designed and managed to optimise the learning benefits?

Recognising 'added value'

Sensors are now available for monitoring virtually every type of physical quantity imaginable; added to the standard laboratory quantities such as temperature, voltage, sound, light and time, there are sensors for measuring force, pressure, motion, acceleration, humidity, pH, oxygen, carbon dioxide, infra-red, and so on. This expansion of opportunity alone adds value to science education, yet the further benefits accruing from the sensitivity and accuracy of many sensors at an affordable cost are a further bonus.

The speed of electronic processing in dataloggers makes possible observations of very short-lived transient phenomena; loggers set up to collect data over long periods of time permit experiments which are inconvenient in the context of school timetables subdividing the day into discrete lessons; and experiments which take only a few minutes are easily repeated to test reliability or to explore the effects of varied conditions. In short, there are many types of datalogging experiment fulfilling a variety of curriculum niches. The selection of examples in this chapter attempts to highlight the most notable features of datalogging and the range of teacher actions which secures learning benefits from them. For ideas on developing a broad datalogging curriculum, teachers are referred to more extensive practical manuals.

2.2 The teacher's role

It is quite natural for teachers newly adopting ICT in their teaching to attempt to adapt the new technology to their existing ways of working. A recent survey (Rogers and Finlayson, 2003) has revealed that for many this means that datalogging is first introduced to students in the manner of a whole class demonstration, with the teacher firmly in control of the experiment and managing students' engagement with the science involved. Many of the same teachers admitted how beneficial it would be for students themselves to get their hands on the equipment and learn first-hand from the experience of controlling and observing the variables in the experiment. Other research (Sandholtz, Ringstaff and Dwyer, 1997) has demonstrated how teachers tend to hand over more responsibility to students as their own confidence with the technology develops and they gain vision of the interactive qualities of learning with ICT.

Even with the traditional pedagogy of whole class teaching, there are several aspects of datalogging methods that challenge teachers to change their thinking about their teaching approach. For example, the graphical display of data provides a qualitative view as a first experience; in contrast with conventional methods in which the readings have to be gathered and tabulated before the graph can be drawn, the graph becomes the starting point rather than end point of class activity so that thinking about the meaning of the graph shape is projected to the forefront of class discussion. Quantitative analysis extracting numerical information may follow, but the order in lesson process is effectively a reversal of the traditional one.

The time scale for a datalogging experiment also has a profound effect on the planning and management of student activity in the lab. Short transient phenomena captured in a few seconds call for activity with a different emphasis from that of the collection of data. The main options here are to spend more time on repeating the experiment, comparing the effects of similar or varied conditions or on detailed analysis of the data. Similarly with real-time experiments lasting up to 5 minutes or so, there is scope for repeating experiments and focusing students' thought on comparing data. For medium-term experiments lasting up to 30 minutes, real time capture is also appropriate, with activity still conveniently fitting into the time constraints of normal lessons. However, in this context, since logging and plotting is automatic, students need to be given tasks to replace the gathering activity normally required in conventional experiments. There are several options:

- to make visual observations of the apparatus or materials involved in the experiment – students can be on the lookout for changes appearing on the graph to see how these might be connected with the phenomenon under investigation

- to intervene in the experiment, introducing a change deliberately, for example stirring a liquid, and observing the effect on the graph
- to make predictions about the graph, or try to explain the graph, or explain what the shape teaches them about the phenomenon
- to answer some questions that link the experiment to previous knowledge or other experiments they have done.

Finally, long-term experiments, lasting hours or days, can only engage students during the planning, setting up and data analysis stages. In these contexts careful thought needs to be given to the most suitable framework for supporting such activities.

As with any type of ICT application, it is necessary to consider the skills requirement for operating the apparatus and software. Skills fall into two main categories: *operational* skills comprising technical competence in setting up and handling the sensors, interface and software, and *application* skills concerned with how the hardware and software tools are put to use in finding information and developing scientific understanding. For acquiring the operational skills, worksheets, manuals and on-screen tutorials are useful aids supporting the training of students. Amongst these skills, those for manipulating graphs feature prominently and a summary of these is presented in Box 1 (page 52). For the application skills, direct teaching is required. Students need to learn how to interpret the shape and features of graphs, how to use data for describing and relating variables, how to make predictions and how to use models. See Newton and Rogers (2001) for a fuller discussion of skills. References to both types of skill will be made in the examples of experiments that follow.

2.3 Experiments using sensors

Experiment with a temperature sensor – 1. Cooling of stearic acid

This experiment is an adaptation of the traditional experiment for measuring the melting point of stearic acid. Crystals of the stearic acid are placed in a clamped boiling tube which is then heated in a water bath maintained at boiling temperature. Heat is applied to the water bath until all the substance is in the liquid state. At this stage the temperature probe is carefully inserted. In order to ensure that the probe is at the same temperature as the liquid, it is necessary to boil the water for another minute, taking care not to scorch the flex connected to the probe. The source of heat is then switched off and the datalogging software set to record data for 20 minutes. Still in the clamp, the boiling tube and its contents are removed from the water bath and the liquid is gently stirred using the probe (Figure 2.6a, overleaf). Stirring ensures that the substance is at a uniform temperature at each stage of the cooling and the resulting graph shape (Figure 2.6b) is more decisive than if left static.

Learning outcomes

This being an example of a 'real-time' experiment, the graph is gradually building up on the screen while students observe the changes taking place in the substance. It is important to prompt students to observe the connection between the appearance of the substance and the graph. The two phases of the substance are associated with distinctive portions of the graph. In particular, the flat part of the graph relates clearly to the formation of white crystals and the increasingly 'sticky' feeling as the probe is gently moved. Discussion can focus on the physical reasons for the flat portion and how ideas about this might be further tested.

Analysing the graph

The graph itself offers several opportunities for exploration and analysis:

1. Dragging a cursor across the graph recreates the sequence of temperature changes with bold visibility in a graphics window, showing the magnitude of readings.
2. Reading the temperature of the plateau region gives the melting point of the substance.
3. The time extent of the plateau, although not significant as an absolute value, can be usefully compared with the time to cool in the liquid phase; this gives a rough significance to the energy stored as latent heat.
4. The curvature of the graph in the liquid phase can be studied with software tools which calculate the gradient at any point of the graph. This shows that the rate of cooling is greatest at the highest temperatures.

Figure 2.6
*Stirring the
cooling liquid,
and the resulting
graph.*

a)

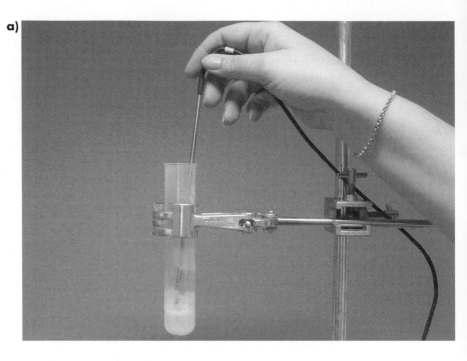

b)

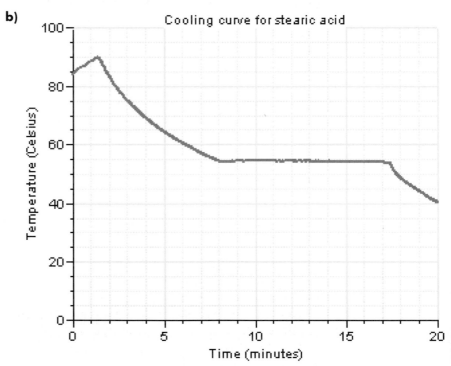

Ideas for investigation

The method described results in a logging activity lasting about 20 minutes, leaving sufficient time in a typical lesson for analysing the graph. Comparing this with the conventional method, the time saved in plotting the graph manually may be spent in thinking about the graph, asking questions about the data and finding out more information from the graph. This is an example of how the automation of the collecting and graphing process facilitates a change of emphasis towards interpreting results. This aspect is supportive of an investigative approach in which students are encouraged to speculate about variables and explore their effect by designing tests under varied conditions. The datalogging system takes care of the routine collection and display and also provides tools for exploring the data.

Here are some ideas for investigations based on the cooling experiment:

1. Try different amounts of substance and note the effect on the length of the plateau.
2. Allow the tube to remain in the water bath as it cools and note how the whole experiment is prolonged.
3. For the cooling phase, place the tube in a bath of cold water and notice how the cooling is accelerated.
4. For the cooling phase, place the tube in a larger tube containing a small quantity of water to create a water jacket. Surround the outer tube with a layer of thermal insulation. Insert a second temperature probe in the water jacket and observe a steady rise in this temperature throughout the experiment and in particular during the plateau phase of the cooling substance.

Other experiments on cooling using temperature probes

- Big and small animals – effect of size on heat loss
- Heat insulation – wrap test tubes in different materials
- Compare wet and dry insulation
- Heat loss and silver foil
- Huddling simulation – effect of clustering test tubes
- Cooling experiments (including hot tank and house insulation etc.)
- Mixing hot and cold liquids
- Radiation absorption/emission
- Thermal conductivity of metals

- Thermal insulation of coffee cups
- Evaporation and cooling
- Freezing and thawing

Experiment with a temperature sensor – 2. Respiration of grass cuttings

This experiment is an example of datalogging over long periods, in this case up to two weeks. The datalogger has three temperature probes connected and is set up to work in 'remote' mode, that is independently from the computer. Freshly mown grass cuttings are heaped into a large bucket or bin. One probe is inserted right in the middle of the heap, a second is placed on top of the heap and a third is placed outside the bucket to monitor the ambient temperature (Figure 2.7a). The datalogger is set to record and is left for about two weeks. During this time respiration causes the temperatures inside and on the heap to rise dramatically. After 12 to 14 days the data is downloaded to the computer and students are then able to view it all on a graph (Figure 2.7b).

Learning outcomes

During an experiment of such length it is only possible to make occasional informal observations of the rise in temperature of the grass cuttings to confirm that the process of respiration is under way. The data collection process would be unviable without a datalogger, so this is a good example of a new opportunity afforded by the technology. It is convenient to set the logger to work in 'remote' mode, leaving the computer free for other tasks during the two week experiment! However, this mode does not permit any interaction with the data until collection is complete, so student activity is confined to analysing data retrospectively. The graph is a rich resource for discussion as students consider the similarities and differences of the three graph lines and seek to explain the shapes in terms of their understanding of respiration and the climatic conditions during the experiment.

Analysing the graph

Observations of the graph and measurements can focus on several features:

1. The temperature difference between the core and the top of the heap.
2. The dramatic rise of temperature during the first day.

Figure 2.7
The set-up with the grass cuttings and temperature probes, and the resulting graph.

a)

b)

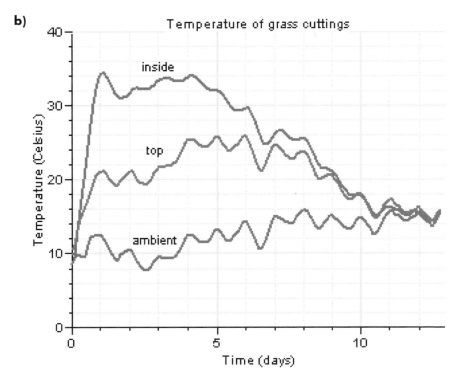

3. The gradual increase in temperature after the first two days.
4. The daily cycle of change associated with the succession of night and day, as shown by the graph of ambient temperature.
5. The oscillating variation of the grass temperatures, indicating how the grass is subject to the daily changes in the local climate.
6. The decrease in grass temperature during the second week. Software tools for fitting curves may be used to test for exponential decay of temperature.

Other investigations of an environment using sensors

- Dawn chorus – remote logging of sound and light
- Global warming – how carbon dioxide effects the warming of air in a plastic bottle
- Day and night – plot temperature, light and humidity over a 24-hour period
- Rainfall measurements
- Daily changes in a habitat
- Air stratification – room temperature and height
- Heat loss through windows
- Loft insulation
- Temperature control in a freezer

Experiment with a light sensor – rate of reaction

The idea for this experiment is taken from the familiar 'disappearing cross' experiment for investigating the rate of reaction between sodium thiosulphate and hydrochloric acid. The reaction produces a dense white precipitate which can be monitored by passing a beam of light through the reactants and measuring the intensity transmitted. The experiment is conducted in a 100 ml beaker in a wrapping of black paper, with a torch light above and a light sensor below (see Figure 2.8a, page 38). 50 ml sodium thiosulphate (1 M) is first poured into the beaker and the datalogging software is set to record the light level for one minute. A syringe is used to squirt 5 ml 1 M hydrochloric acid into the beaker. In moderation, the squirting action is effective in mixing the solutions well. It also causes a short dip in the light readings which usefully marks the time at which mixing occurs.

The experiment is easily repeated using the same quantities of solution and acid but with the sodium thiosulphate diluted to 50% of its initial concentration, i.e. 0.5 M. The software should be arranged to overlay the second set of results on the same graph for easy comparison (see Figure 2.8b).

Learning outcomes

The shape of the graph reveals that the rate of reaction varies considerably during the experiment; the initial period of no apparent change is followed by a period of rapid change which then degenerates into a 'tail' of slow change. The 'S' shaped curve prompts a discussion of the need for a criterion for evaluating the rate of change. The gradient of the graph is a useful indicator and is easily calculated with software, but a judgement has to be made about which part of the curve is the most appropriate to measure. Applying the same criterion to both sets of results allows fair comparisons to be made and the effect of reduced concentration causing a slower reaction may be deduced. The shape of the tail is explained by considering the change in the concentration of the reactants as the reaction proceeds.

Analysing the graph

The graph offers several opportunities for exploration and analysis:

1. The graph shape has a number of interesting features evident from simple inspection: the dip at the beginning marking when the acid was introduced; a period of no change; a steep fall; the extended tail.
2. The curved portion of the graph can be studied with software tools which calculate the gradient at any point of the graph. This shows that the rate of change is initially high but gradually reduces.
3. Time intervals from the moment of mixing may be measured to find the duration of the flat portion, the time to reach the maximum rate and the duration of the tail.
4. The first derivative of the curve is useful for identifying the point of maximum gradient.

Figure 2.8
The apparatus for investigating rate of reaction, and the resulting graph.

a)

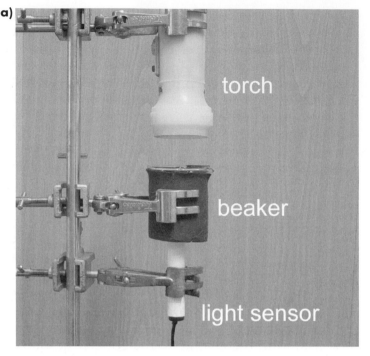

b)

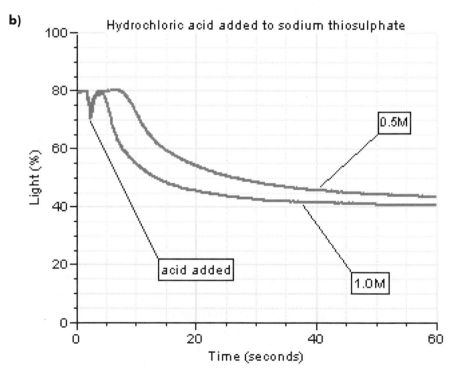

Ideas for investigation

Again, since the experiment is conveniently repeated and new graphs obtained with ease, the datalogging method encourages further investigation of the variables involved:

1. The basic investigation into the effect of concentration on rate of reaction may be refined by further repetitions of the experiment with a graduated series of dilutions, yielding a family of curves. Plotting a further graph of the maximum gradient against concentration gives a clear indication of the trend.
2. The effect of temperature may be investigated by placing the beaker in a larger beaker containing warm water and repeating the experiment at several different temperatures. The results may be evaluated by plotting a graph of the maximum gradient against temperature.

Experiment with a pH sensor – effect of antacid tablets

This is a simple experiment to demonstrate the effect of indigestion tablets (e.g. Alka-Seltzer) on a weak acid. Similar quantities of cola drink are placed in two beakers so that each is half full. A pH probe is placed in one beaker (Figure 2.9a) and the datalogging software is set to record for 4 or 5 minutes. An Alka-Seltzer tablet is inserted into the drink. As it dissolves, the pH value will be observed to change. The experiment is then repeated in the second beaker, this time crushing the tablet before it is inserted. The software should be arranged to overlay the second set of results on the same graph for easy comparison (see Figure 2.9b).

Learning outcomes

The graphs show the change of acidity of the cola as the tablet dissolves. The overall change, time for change and rate of change may be obtained from the graph and used to discuss the usefulness of the tablets for curing indigestion. The merit of crushing the tablet, i.e. the effect of particle size on the speed of dissolving, may also be discussed.

Analysing the graph

With two sets of results overlaid on the same graph axes, it is easy to compare the two versions of the experiment and obtain useful measurements.

Figure 2.9
The set-up with the pH probe and the resulting graph.

a)

b)

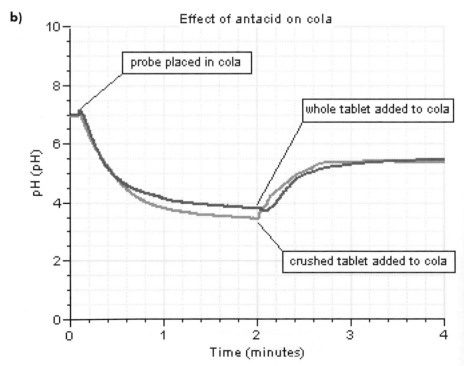

1. Inspection of the graphs makes it very clear that the pH value changes and that the time for change is less in the case of the crushed tablet.
2. Software tools may be used to measure directly the time period for the change and the average rate of change in each case.

Ideas for investigation

The datalogging method allows further investigation, for example:

1. How does the addition of a second tablet affect the pH of the solution?
2. Try dissolving the tablet in water first and adding the solution to the cola.
3. Repeat the experiment with different types of drink to find out the effectiveness of the tablets more generally.

Other investigations using a pH probe

- Souring of milk
- Titration curves
- Acid rain: pH of rainwater
- Enzymes and temperature

Experiment with a humidity sensor – transpiration of leaves

This experiment uses very simple resources and is conveniently left to run during a lesson period in parallel with other activities. A freshly picked leaf is placed inside a plastic bag together with a humidity sensor (Figure 2.10a). The bag is sealed and the datalogging software is set to run for 20 to 30 minutes. Since there is very little happening that is visible, it may be as well to set up the logger in 'remote' mode and download the data after the set time has expired.

Learning outcomes

If the bag is sealed effectively, the humidity rises to a steady value which may be compared with the initial or ambient humidity. See Figure 2.10b. The factors leading to an equilibrium in the humidity can be discussed. Observation of the effect of breaking the seal on the bag can stimulate thinking about diffusion.

Figure 2.10
The set-up with the humidity sensor, and the resulting graph.

a)

b)

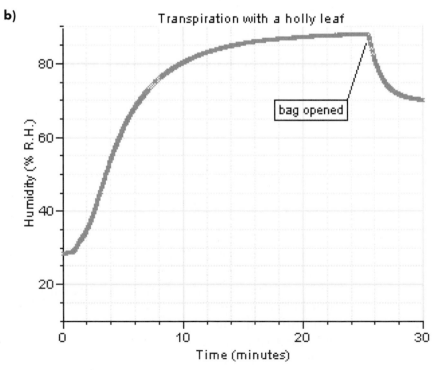

Transpiration with a holly leaf

bag opened

Analysing the graph

The experiment is best suited to qualitative interpretation, so the role of the graph is to display change and to prompt thinking about the processes occurring inside the bag, rather than numerical measurement. It can be observed that the rate of change diminishes as equilibrium is established.

Ideas for investigation

The datalogging method allows further investigation, for example:

1. Repeat the experiment with a leaf still attached to a well-watered growing plant and compare the graph with that for the detached leaf.
2. Try the experiment with different types of leaf and consider how some types of leaf are better than others at retaining their water content.
3. Make a model leaf out of filter paper by drawing round the leaf on the paper and cutting out the leaf shape. Soak the model leaf in water and repeat the experiment to compare the model with the real leaf. Consider the similarities and differences between evaporation and transpiration.

Other investigations on plants using a humidity sensor

- Photosynthesis
- Respiration of peas

Experiment with a motion sensor – a trolley on a slope

This experiment is one of many that use a motion sensor for continuous monitoring of position or distance. The sensor is placed near the top of an inclined runway. At the opposite end, down the slope, is a solid barrier. A standard dynamics trolley is placed at the top of the runway with its spring-loaded plunger pointing down the runway. See Figure 2.11a. The trolley is allowed to roll down the runway and the plunger acts as an elastic buffer, making the trolley bounce off when it collides with the barrier. After the first collision, the trolley moves back up the runway, slowing down all the time, and eventually stops and rolls down again. The cycle repeats several times until the trolley finally comes to rest. The whole motion is recorded in real-time by the datalogger.

Figure 2.11
*The set-up with
the motion sensor,
and the resulting
graph.*

a)

b)

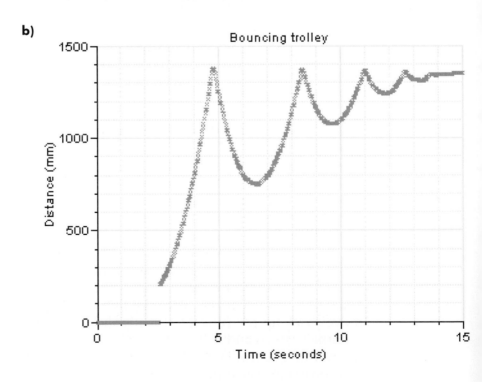

Learning outcomes

The real-time plotting of results on a distance–time graph with the motion sensor is a graphic example of datalogging offering a totally new experimental opportunity. The closest equivalent conventional activity is with tape and a ticker timer, but even this cannot cope with the reversals in the motion of the trolley. The connection between the observed motion and the resulting graph (Figure 2.11b) is immediately apparent, although students need to be careful to observe that distance measurements are from the sensor at the top of the runway rather than from the barrier at the bottom. The parabolic appearance of the graph can be associated with acceleration and deceleration. The reversal points on the graph can be associated with the collisions. The decay of the oscillations can prompt thinking about the type of collision with the barrier. The asymmetry of the graph segments can prompt thinking about friction.

Software facilities for calculating and plotting further graphs of velocity–time and acceleration–time can be the basis for an even more sophisticated exploration of the motion of the trolley.

Analysing the graph

The graph provides a rich source of exploratory activity:

1. Dragging a cursor across the graph simulates in a graphics window the oscillating motion of the trolley up and down the runway.
2. The velocity before and after each bounce may be measured and compared using a software gradient calculation tool.
3. The distance the trolley moves up the runway after each bounce may be measured and compared for successive oscillations.
4. The time intervals between successive collisions may be compared.
5. Software can calculate and plot data for a velocity–time graph which may be compared with the original distance–time graph.
6. The slope of the velocity–time graph indicates the acceleration of the trolley at that point in the motion. It is particularly interesting to compare accelerations up and down the runway, before and after each collision.

Ideas for investigation

Although the practical arrangement contains only a few simple components, there are numerous extensions to the basic experiment. Some possibilities are:

1. Find out how the slope of the runway affects the graph.
2. Find out the effect of increasing the mass of the trolley, by attaching further masses.
3. Add a wind brake to the trolley to increase air friction.
4. Retract the spring buffer and replace it with a cardboard 'crumple zone'.

Other investigations using a motion sensor

- Measurements of human motion
- Oscillations (mass-on-spring and pendulum)
- Elastic and inelastic collisions

Experiment with a light gate sensor – force and acceleration

This experiment illustrates a technique entirely different from that of the motion sensor for measuring the motion of objects. It is based upon earlier conventional experiments with light gates and electronic timers, whereby the object under scrutiny passes through a light beam, interrupts the light falling on a sensor and thus initiates a time interval measurement. The speed of the object may be calculated from the length of the object and the interruption time interval. The principle of the datalogging version is the same, except that the computer both measures the interruption time interval and calculates speed from information supplied about the length of the object. To measure acceleration, the object is fitted with a double segment card (see Figure 2.12a) so that the software can measure two velocities in quick succession and calculate the rate of change of velocity.

In this experiment the datalogging system measures the acceleration of a trolley resulting from a constant force applied via a piece of string pulled by a falling mass. Unlike the previous experiments which involved a rapid sequence of measurements, here each run of the trolley yields only a single measurement of acceleration. However, it only takes a few moments to reset the trolley and have another run, so, through repetition, a series of measurements is possible. For the forces proposed here, it is desirable to make the mass of

Figure 2.12
The set-up with the light gate, and the resulting graph of acceleration v. force.

a)

b)

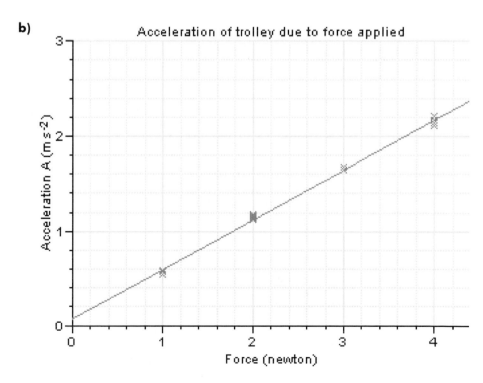

the trolley 2 kg. The value for the force acting on the trolley must be entered via the keyboard. Starting with a mass of 100 g, giving rise to a force of 1 N, a series of measurements is recorded. This is repeated for masses of 200 g, 300 g and 400 g, corresponding to forces of 2 N, 3 N and 4 N. The software is configured to plot a graph of acceleration against force (Figure 2.12b).

Learning outcomes

The advantage of discrete measurements with light gates is that the calculation process for velocity and acceleration is transparent and, with suitable teacher guidance, reinforces understanding of their definitions. The fact that software takes care of the measurement and calculation process allows students in this case to focus their attention on the relationship between the force and the acceleration. This is a welcome advance on older methods which require considerable amounts of manual calculation effort. Conclusions about the relationship between force and acceleration are entirely dependent on successful analysis of the graph. A straight line graph verifies Newton's second law of motion which states that the acceleration of an object varies in proportion with the resultant force on the object.

Analysing the graph

Unlike graphs produced in the previous experiments, which showed a range of values spread uniformly across the horizontal axis, the graph in this experiment shows four clusters of points, one for each value of force used in the experiment. The vertical spread of each cluster shows some variability in the acceleration measurements, but usually this is very small compared with the changes resulting from increasing the force.

In view of the 'low density' of points, this type of graph calls for a different treatment when attempting to investigate the relationship between acceleration and force. In this case it is appropriate to employ the 'best fit' tool in the software to calculate and plot the best line through as many points as possible. Such a line identifies the trend associated with the distribution of points and bridges the gaps between the points.

It is usual to find that the line, although straight, does not pass through the origin and students may be prompted to think about two factors by way of explanation.

1. What role does the force of friction play in the experiment? How might the results be affected by friction?
2. Was the same mass accelerated each time a measurement was recorded? It is easy to forget that the falling mass experiences the same acceleration as the trolley and the amount of falling mass is varied.

Ideas for investigation

The conditions in the experiment may be modified to explore the effects of friction and of changing the mass of the trolley. The traditional method of compensating for friction is to use a slightly sloping runway.

Other investigations using a light gate sensor

- Predicting the motion of a falling card: dependence of velocity on height fallen
- Measuring the speed and acceleration of a model car
- Pendulum measurements
- Potential and kinetic energy

Experiment with voltage and current sensors – electrical properties

This experiment requires a simple electric circuit containing the following components connected in series: resistor, battery, variable resistor and current sensor. Further to this, a voltage sensor is connected in parallel across the resistor which is the component under test. See Figure 2.13a. The datalogging software is set up to show a graph of voltage and current plotted against time. Data is recorded for half a minute or so, during which the variable resistor is adjusted continually to produce a change of voltage across the resistor. It is instructive to vary the voltage across the available range, making it increase, decrease and pause during recording. Students should observe and compare the changes in the current and voltage readings using the graphical presentation aids provided in the software.

Learning outcomes

With this type of experiment, datalogging breathes new life into investigations of the electrical properties of resistors and other circuit components. Simultaneous measurements and graphical presentation of current and voltage are exploited to demonstrate the relationship between these electrical

Figure 2.13
*The circuit and
the resulting
graph.*

a)

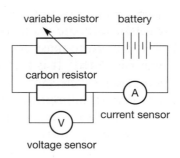

b)

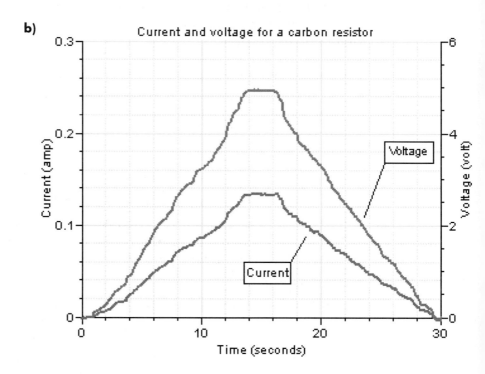

quantities, and software calculation tools permit easy access to
the derived quantities of resistance and power. The suggested
graph display against time is unconventional for typical
voltage–current investigations, but it has the advantage of
showing clearly how the voltage and current vary in unison for
a simple resistor (see Figure 2.13b). This gives a direct
appreciation of the essence of Ohm's law which states that the
current varies in proportion to the voltage. The proportionality
can be readily tested using cursors to read off corresponding
values of voltage and current.

The axes can then be redefined to give the more conventional current v. voltage graph. The linearity will be clear and the gradient can be recorded. The latter gives an indication of the 'conductance' of the resistor, or 1/resistance, which is clearly constant across the whole graph. The next graph to plot is one of resistance v. voltage; software makes it easy to calculate the resistance from pairs of voltage and current readings. The resulting line is horizontal, confirming that the resistance is indeed constant for the resistor, an ohmic conductor. Finally a graph of power v. voltage may be plotted, this time showing a parabola which may be tested for its quadratic property.

Analysing the graphs

This experiment illustrates well how several different graphs may be derived from a common set of data. Each graph has a story to tell, assisted by the software tools:

Voltage and current v. time

Dragging a cursor across the graph recreates the sequence of voltage and current changes with bold visibility in a graphics window, showing the magnitude of readings. It is seen how the rises and falls in the current exactly follow those of the voltage. The cursor readouts can demonstrate how the current doubles whenever the voltage doubles, and so on.

Current v. voltage

The straight line demonstrates proportionality which may be tested using the cursor to read off current values when the voltage is, say, 0.5, 1.0, 1.5, 2.0 V and so on. The current is observed to increase in the same ratio as the voltage. Assuming voltage, as the independent variable, to be plotted horizontally, the gradient indicates how good the resistor is at conducting electricity. Thus a steeper graph is expected for a better conductor (of lower resistance). This is a useful discussion point when comparing graphs for different resistors. The software provides a direct measure of the gradient and the resistance value can be found by calculating the reciprocal of the gradient.

Resistance v. voltage

This yields another straight line graph, this time parallel to the voltage axis, confirming that the resistance does not change as the voltage is varied.

Power v. voltage

Taking readings at regular intervals of voltage shows that the power increases in proportion to the square of voltage. This can be further verified by using a curve-fitting software tool or by plotting a further graph of power v. voltage squared.

Ideas for investigation

In view of the fact that a set of raw data and a graph for current and voltage may be obtained in less than one minute, it is convenient to repeat the experiment with different components:

1. Try several resistors of different values and compare the graphs obtained.
2. Combine two similar resistors first in series, then in parallel and compare the results to deduce the effective combined resistance.
3. Try the experiment with a torch lamp, first increasing the voltage steadily, then decreasing it. Notice that the current v. voltage graph when voltage is increasing is different from that for when it decreases, and that both are non-linear. Discuss the effect of temperature on the resistance of the bulb.

Other investigations using electrical sensors

- Investigating a bicycle dynamo
- Investigating a solar cell
- Switch bounce measurements
- Capacitor charge/discharge experiments
- Electrical induction
- Magnetic field strength in a coil

Box 1. Learning from graphs

Shape

The shape of a graph gives much useful qualitative information about the events in an experiment: upward and downward trends clearly indicate increases and decreases in quantities, and horizontal lines indicate steady conditions. Rapid or gradual changes show up as steep or shallow graph lines. Maxima and minima often indicate significant events or transitions. To observe such features and make comparisons between different graphs or parts of

---

graphs is the first step in interpreting the data represented in graphs. However, it is important to take account of the effect of the plotting scale on the magnitudes of such changes and rates of change, since the scale can easily magnify the appearance.

Measurement

Data *values* for any point on a graph line may be read directly by manoeuvring cursors in the graph area. A facility for locking the cursors on to data points ensures complete accuracy. Further tools use pairs of cursors to calculate *differences* and *intervals* between vertical and horizontal coordinates, the *gradient* at any point, the *rate of change* between two points, the *ratio* between two readings and the *area* underneath the graph line. Using the variety of possible measurements, students learn that graphs contain much more information than first meets the eye.

Processing

Real data sometimes has an untidy appearance due to the effect of hidden or unwanted variables. In such cases it is often more useful to seek an underlying trend in the data rather than study the minute detail. Software tools for removing the 'jerkiness' in data conveniently reveal patterns as smoother lines which can help to establish relationships.

Certain physical quantities such as 'energy' and 'acceleration' depend upon more than one primary variable and have to be derived by calculation. Software provides various means of automating this process and displaying derived quantities directly on a graph.

Curve fitting

Observing the shape of a graph is important for giving a first indication of a trend or relationship, but a mathematical formula adds precision to the process of describing shape and defining its characteristics. Traditionally the straight line graph is used in the quantitative analysis of data, and students are taught the significance of the gradient and intercept and to use the graph as an instrument for predicting data. All this and more is achievable with great facility using curve-fitting tools which offer families of common curves for fitting as well as straight lines.

Box 2. Thinking about sensors

The sensors used in datalogging are essentially electrical devices, each yielding an electrical signal which indicates the magnitude of a physical quantity. A wide variety of sensors are available, covering most physical measurements encountered in school science. Broadly, there are two types of sensor, the switch ('digital') type providing the on/off signals used in experiments involving timing and frequency measurements, and the analogue type offering continuously variable signals. When choosing sensors it is important to consider the different properties of the two types so that the output signal is suitable for the application. In many cases the distinction is obvious but ambiguity sometimes exists with sensors for measuring radioactivity (pulses or pulse rate), heart activity (beats or beat rate) and light (intensity or change).

For every sensor there is a response time which determines how promptly the sensors can signify a change. For light gates the response time needs to be very fast indeed. For temperature measurement, because most types of probe possess thermal inertia, it is necessary to think about how long to wait for readings to settle down when changes of temperature occur; unfortunately, as the difference in temperature between the probe and its surroundings reduces, the rate of response of the probe also reduces. pH probes also require a settling time when inserted into a solution. Sound level sensors usually have an internal hysteresis to reduce jerkiness in readings but this can also limit their speed of response.

Many sensors are capable of great sensitivity to small changes but this can sometimes be a distraction when values are displayed with too many decimal places or a graph-plotting scale produces excessive magnification. In either case it is wise to adjust the display. Sensitivity should not be confused with the accuracy of readings, which depends upon the quality of calibration of the sensor. Most analogue sensors are pre-calibrated but some types of sensor such as pH probes and angle/position sensors require calibration at each time of use. Inexpensive sensors, such as light sensors, are sometimes uncalibrated. They are mainly useful for observing changes and making rough comparisons, but quantitative conclusions should be avoided. Oxygen, pH and conductivity sensors cannot generally be used in combination because they interact with each other and give false readings when placed in the same solution.

2.4 Links with modelling

A promising area of activity involves bringing theory and practical work together, whereby real data collected from experiments using dataloggers is compared with 'synthetic' data generated by mathematical models. The construction of a model in software requires students to apply their previous knowledge and understanding of a phenomenon (see Chapter 6). The opportunity to validate a model against the equivalent real experiment supports both thinking about models and the interpretation of collected data. The electrical experiment described above (pages 49–52) provides a useful illustration of this: the collected data may be compared with data calculated by a model based on Ohm's law. If the correlation between the two sets of data is good, this helps to validate Ohm's law as a description of the behaviour of the resistor.

Another use of models is to extend an investigation beyond the limitations of normal laboratory conditions. In an investigation using real equipment and materials, there is a natural limit to the extent to which variables and parameters may be controlled. However, a model simulating the same phenomenon need not have the same constraints; values may be assigned to variables at will in the virtual world created by modelling software. For example, a model may be created to simulate the experiment on evaporation described on pages 19–20 (Figure 2.1), such that students may explore the effect of ambient temperature by specifying any value they wish, or to simulate the behaviour of many different liquids by varying the latent heat of vaporisation or volatility factors.

2.5 Managing datalogging in the classroom

As with other applications of ICT, the learning benefits of datalogging are best realised when students themselves are allowed to handle the equipment and interact with the software. However, limited amounts of equipment usually mean that the logistics of whole class participation simultaneously, in the same manner as a traditional class practical session, is unrealistic in most school situations. The most common solutions adopted by teachers to overcome this limitation and to optimise participation are as follows:

1. When only one set of datalogging equipment is available, it may be used in parallel with other group activity using the equivalent conventional method. Students' results may be compared with the datalogging results, and the qualities of new and old methods discussed.
2. A single datalogging experiment may be one of a number of different experiments organised as a circus, or available as an ancillary activity servicing students' paper-based work or conventional practical work.
3. When the datalogging experiment is quick, students' work may be organised in a rota to give each group an opportunity to obtain their own results.
4. Dataloggers may be used in the science lab to collect the data and then students transfer to the IT suite to download the data and analyse it on the computers. Variants of this approach include splitting the class in two halves so that a shift system of working is employed, or spreading the collecting and analysing activities over two lessons.

◆ *Conclusion*

To conclude this chapter we will return to the question of the appropriate use of datalogging. The example experiments have illustrated numerous features of datalogging methods which can add value to students' experience of practical science. These form a strong basis for identifying appropriate use, especially when the feature is not replicated in conventional methods. The following list summarises the principal reasons for choosing datalogging methods:

♦ Datalogging is most useful when used to observe and measure *changes*. Students can 'see' changes and rates of change easily as graphs and graphics displays. For one-off measurements, conventional instruments are likely to be more appropriate.

♦ Datalogging offers forms of measurement which extend our senses, facilitating very fast data-gathering, or data-gathering over long periods of time.

♦ Sensors can offer superior sensitivity, precision and accuracy over conventional instruments (see Box 2, page 54, for some qualifying comments).

♦ Real-time interaction supports an investigative approach and helps students to make connections between a phenomenon and its graphical representation.

♦ The plotting of graphs is subject to much less error compared with manual methods and students can experiment to learn about the effect of the axes' scales on the appearance of a graph.

♦ A qualitative view of data is available without having to labour with excessive quantities of numbers.

♦ The quantitative handling of data offers accuracy and sophistication, due to the availability of labour-saving tools for analysing the data and calculation facilities permitting the derivation of new data.

♦ Data sets of simultaneous or time-stepped measurements are easily compared.

An overarching criterion for appropriateness is that the qualities offered by datalogging fulfil learning objectives. It is often necessary to re-appraise learning objectives when novel features of datalogging displace conventional methods or create new opportunities. This particularly applies when graphs are involved; clearly, objectives associated with the skills required for and understanding gained from conventional pencil-and-paper techniques are not served well by software; however, the range of ideas presented here for analysing and manipulating graphs indicates objectives for which tasks would otherwise be tedious, time-consuming or impossible without software. The justification for such tasks rests upon their usefulness for helping students to investigate data, to discover patterns and relationships which might otherwise be hidden.

Finally, we return to a theme which is implicit throughout this book, that the realisation of learning objectives is not achievable by the unaided use of software but is underpinned by the decisions and actions of teachers. Our examples have illustrated potential learning benefits, but to convert the 'potential' into 'actual' benefit, the teacher has a vital role, through the planning of activity, involving definition of objectives and identification of training needs, and the conduct of lessons, involving numerous types of interactions with students. In both respects, planning and management, much conventional teaching wisdom, experience and skill still apply; for example, the logistics of practical work in small groups, strategies for an investigative approach, target setting, making links with previous knowledge, assessing students' understanding, are familiar concerns for science teachers. We hope that the discussion presented in this chapter prompts thinking about how these might be adapted when datalogging is introduced.

◆ *References*

Newton, L. R. and Rogers, L. T. (2001) *Teaching Science with ICT* London, Continuum.

Rogers, L. T. and Finlayson, H. M. (2003) *Does ICT in science really work in the classroom? Part 1: The individual teacher experience* School Science Review June 2003 84(309).

Sandholz, J. H., Ringstaff, C. & Dwyer, C. D. (1997) *Teaching with Technology: Creating student centred classrooms* New York, Teachers College Press.

Wellington, J. J. (1998) *Practical Work in School Science – Which way now?* London, Routledge.

◆ *Resources*

National Curriculum

The Programmes of Study for Science contain margin notes on opportunities for ICT use, including datalogging. www.nc.uk.net

Datalogging Insight

This software (the fourth version of *Insight*) is unique in combining diverse graphing tools, modelling facilities and compatibility with datalogging hardware from all leading UK manufacturers. Supplied with a range of worksheets for datalogging and modelling.

Logotron Ltd, 124 Cambridge Science Park, Milton Road, Cambridge CB4 0ZS

www.logo.com

Roger Frost's *Dataloggerama*

This website contains a comprehensive range of resources for supporting datalogging activities; advice on choosing equipment, ideas for experiments, data file downloads, training facilities and so on.

www.rogerfrost.com

ICT Activities for Science (1998–99) Heinemann

Worksheets for datalogging.

www.heinemann.co.uk

Data Harvest

Manufacturers of the *EasySense* range of dataloggers and *SmartQ* sensors. Publishers of curriculum packs supporting datalogging.

www.data-harvest.co.uk

DCP Microdevelopments

Manufacturers of the *LogIT* range of dataloggers and sensors.

www.dcpmicro.com

Philip Harris Education

Manufacturers of dataloggers and sensors. Publishers of curriculum packs supporting datalogging.

www.philipharris.co.uk

Pico Technology

Manufacturers of a range of inexpensive datalogging equipment. The website offers a comprehensive range of ideas for experiments.

www.picotech.com/experiments

3 Using multimedia

David Sang

◆ Introduction

The term 'multimedia' has a range of meanings. In this chapter, it is used to mean material that generally combines text, graphics, video and sound in an integrated package. Such software packages are made available either on CD-ROMs or via the World Wide Web.

3.1 Why use multimedia?

We will start by outlining some of the merits of using multimedia in science teaching. Later in the chapter, you will find these points spelled out in greater detail in the context of the specific examples considered there.

- Multimedia can give students **rapid feedback**. They click on something and they are told whether their choice is right or wrong. They test an idea and see the results within seconds. This is highly motivating – it's why people will play with computer games for hours on end.
- Multimedia can have great **ease of access**. With a CD-ROM in your bag or a Java applet (see page 72) downloaded onto your laptop, you can be sure that the items you want to use in a lesson are immediately available. In addition, access to a library of multimedia materials will allow you and your students to move rapidly between items in a way which is incomparably faster than searching in a conventional library.
- Multimedia can **show motion**. Books and magazines have static images. Multimedia can incorporate video clips, and this is much more convenient than searching through a cassette tape for the appropriate part of a recorded TV programme. It can also include animated images which will upstage most static images in a textbook.
- Multimedia can **show the invisible**. An understanding of atoms and molecules is at the heart of chemistry and much of physics, and there are many multimedia packages which will allow your students to visualise how the atoms and

molecules move and reorganise themselves as chemicals react. But microscopic particles are not the only 'invisible' things that we expect our students to visualise. Forces, for example, are also invisible. We can represent them with arrows, showing their magnitude and direction and how these change. Magnetic, electric and gravitational fields are also invisible, and so on.

- Multimedia can **change speeds**. Things that are too slow to be conveniently observed can be speeded up, for example the growth of a plant, the motion of tectonic plates, the rusting of a nail. Things that change too quickly can be slowed down, for example the motion of a bouncing ball, the reaction of caesium with water. A tedious, repetitive experiment can be run in simulation many times during a single lesson.

- Multimedia can enable students to answer their **"What if?"** questions. A simulation can allow students to alter variables or change configurations so that they can make their own decisions about what to try next. This can be greatly empowering for them.

- Multimedia can help you to **avoid hazards**. Your students can safely investigate radioactive materials, explosive chemicals or harmful micro-organisms when they are working in the virtual world of multimedia.

All of the above may seem too good to be true. How did we ever achieve anything before multimedia came along? Of course, not every multimedia package can deliver all of the benefits listed above. There is good multimedia and there is bad. Worthwhile multimedia has only been around for a few years; it is a technology that is developing fast, and the best examples from five years ago are now likely to look a bit faded. (We shouldn't forget that our students are probably familiar with the most sophisticated software through their use of computer games and their television consumption.)

So, as with any teaching resource, you must assess each item of multimedia on its merits. Will it do something for you which you want to do, and which you cannot more simply achieve by some other means? In the examples that follow, you can look out for suggested reasons why multimedia may be the preferred choice for solving a particular educational problem.

3.2 Multimedia information sources

We want our students to learn about science, but they also need to learn about multimedia and ICT in general. This means that you need to help them develop their ICT skills alongside their understanding of scientific concepts, but you don't want the ICT to dominate the lesson.

A good place to start is by using multimedia information sources. These include encyclopaedias on CD-ROM (for example, *Encarta* and *Encyclopaedia Britannica*). There are also many other information sources covering narrower areas, and these can be more useful as your students are less likely to stray from the area you wish them to consider.

Students will be excited by the use of multimedia; however, you will want to ensure that your lesson is focused on science, so you will need to devise a lesson into which the ICT is integrated. Here are three examples, useful in areas of science that are commonly studied in lower secondary classes.

Astronomy

The aim is for students to learn about the planets of the Solar System. The multimedia tool is *Planet 10*, available on the accompanying CD-ROM, or from the Planet Science website (www.planet-science.com). Other similar packages are available elsewhere, on CD-ROM or websites.

Start by asking students to list everything they can think of that is part of the Solar System. Why is it called the Solar System?

Now use *Planet 10* to show the planets in orbit around the Sun. Zoom in and out, show the planets' orbits, and view them from different perspectives. Discuss such questions as how the brightness of sunlight and the temperature will vary from one planet to another.

In the next stage, students have to research the different planets. *Planet 10* has information pages about each, plus data. Give them a task. It might be to answer the question, "Where would you like to send an exploratory mission, and why?". A frequently used alternative is to ask students to devise a travel brochure for future tourists to their selected planet. Promoting winter sports holidays on a frozen gas giant can be a demanding activity!

Students should be able to copy and paste both text and images; they should rearrange it and add to it, to satisfy the task they have been set. Then they can share their ideas with the rest of the class.

Lessons from *Planet 10*

An activity like this will teach your students about the Solar System in an entertaining way. They will also learn about an important aspect of ICT.

Multimedia resources can store a large amount of information, but you may need to put some effort into finding it. You may have to click on tabs and buttons. You may have to move from one page to the next, or click on links between different pages.

A well designed multimedia resource will have a clear layout, so that the functions of the various buttons etc. are self-evident. A poorly designed resource will lead your students off down a rabbit warren of interconnected pages and links, leaving them unable to get back to the place they started from. Consequently, they will lose sight of the task you have set them.

Muscles, bones and joints

We each have a body, though some bodies work better than others. However, we aren't automatically aware of what goes on inside our bodies. Using a multimedia encyclopaedia such as Dorling Kindersley's *The Ultimate Human Body* you can show the workings of muscles, bones and joints.

The package is far more extensive than you might need for teaching about the mechanics of the body, so you need to be selective. You can show video sequences, use a virtual X-ray scanner, and interact with 2D and 3D models. Suppose you want your students to learn that muscles occur in opposing pairs, because they can pull but cannot push. Ask students to bend their arm at the elbow, so that their lower arm moves up and down. Can they say which muscles they are using? Ask them to draw a diagram of muscles and bones, to show how they think this might work. Then use the multimedia to test their ideas.

Lessons from *The Ultimate Human Body*

Firstly, be selective. A resource like this is designed to be all-encompassing. If students are left to their own devices, they may wander far from any task you set them.

Secondly, the resource is not a lesson in itself; it hasn't been designed in that way. You will have to build the resource into a lesson. That takes time and thought, but it is worth it in the long run. Your students will enjoy the lesson and find it memorable, and you will look forward to the next time you use it.

Chemistry

Many publishers now provide CD-ROMs to accompany their latest textbooks. At the same time, multimedia publishers are starting to produce packages that cover a complete course, and which bring together a mixture of artwork, photographs, text, animations and exercises. These may be available online or on CD-ROM. An example is *Absorb Chemistry for GCSE* by Lawrie Ryan, published by Crocodile Clips. Students can work their way through a sequence of 'pages', reading text, looking at images and using animations. They can test themselves using the various activities which are checked instantaneously by the software. There are even 'investigative' activities in which students can explore the effect of changing a range of variables.

Lessons from *Absorb Chemistry*

This product is rather like an on-screen, interactive textbook. It gives your students a chance to work at their own pace, so that they can skip material with which they are already familiar, or revise difficult topics. They get instant feedback in the exercises, and can repeat tasks until they are happy with the results.

Because students are familiar with textbooks and worksheets, they will have little difficulty in tackling multimedia materials like this. However, you might use *Absorb Chemistry* in a different way. You could project selected items for the whole class to see, and use them as the basis for discussion. Ask students to vote on the correct answer to a question, and test the consensus. Use an investigative tool, and pose "What if?" questions. This gets students into a 'thinking-ahead' mode, so that they are not simply taking in material presented to them, but also thinking about its implications.

3.3 Simulations and models

In this section, I have chosen some popular examples of software which illustrate many of the benefits of using multimedia. They are published by a range of multimedia companies; you may find similar products from different companies.

The terms 'animation', 'simulation' and 'model' are not particularly well defined. Each is likely to show how something changes, so that the view on the screen changes with time. A model has a rather specific sense: it implies that the programmer who has developed the software has used equations that reflect the underlying science. A car may be shown moving according to the equations of motion; particles move at random and collide according to Newton's laws, and so on. An animation or a simulation may be simpler; it might be programmed to give a realistic representation of reality, without necessarily following scientific laws. A model is likely to have more possibilities for flexible use, because changing any parameter will produce a correct outcome.

Food webs

This is a topic that is usually covered early in secondary school, but may be re-visited at a later stage. A useful package for tackling this is *Australian Woodlands* from Newbyte Educational Software.

Students are challenged to create their own food chains and webs. They select a number of organisms from a menu; pictures of each appear on the screen. Then they build a chain by selecting two organisms, the first of which is consumed by the second. (This is fun in itself, as you can select a less familiar animal such as a quoll and try to deduce from its appearance what it might eat.) The chain soon becomes a web.

In the next phase, students can see how populations of the species they have chosen might change over a number of years. This is shown graphically, and almost any web will show interesting patterns: seasonal cycles, predators declining for want of prey, and so on. The software decides for itself reasonable values for the numbers of each species, but you can change these for yourself. For example, you could have a large initial population of wedge-tailed eagles which nose-dives because only small numbers of mammals and reptiles are available.

Taking things further, students can look at the addition of external influences such as drought, burning by humans, or the introduction of alien species such as rabbits or foxes. These are all significant factors in Australian ecology, and you might find video or newspaper extracts to add to the reality of it all.

Lessons from *Australian Woodlands*

When students are constructing a food chain or web, a black arrow links two correctly related organisms. Incorrect answers appear as red arrows, and must be deleted. This shows an important benefit of multimedia: students get instant feedback on their answers, and it is usually clear how any mistake can be corrected. In a typical lesson, the teacher might respond to an individual once or twice; using software, students can get hundreds of responses during the lesson.

The software draws graphs to show how populations vary over time. This is something that electronic systems are very good at. For students to generate such graphs from data is time-consuming and tedious, and they soon lose sight of the purpose of the activity. Instant graphs attract attention, and students can focus on their interpretation.

Finally, it is worth pointing out that the graphs are generated by a true model of the food web. The software 'knows' about the feeding relationships of the organisms, and uses this to work out how the populations will change with time. Because it is a model, and not simply a presentation of a limited set of examples, the user has the freedom to make many different choices, and even to change the scenario part way through. Many questions become possible: "Which species will do better if there is a drought? Could foxes be controlled by encouraging eagles and dingoes?" and so on. The developers of the software may not have thought of the questions that your class may ask, but the software can still answer them.

The seasons

The explanation for the existence of seasons also appears early in the secondary science curriculum (and may be dealt with in the primary curriculum). The idea may seem simple, but it requires students to radically change their viewpoint: from standing on an apparently flat, stationary Earth, to looking down at a spherical, rotating Earth in orbit around the Sun. This sounds like a task for multimedia.

PLATO Learning's *Earth and Sun* is designed to convey the scientific explanations of night and day, and the seasons. It allows you to show, simultaneously, views of the sky from two points on the Earth, the rotating Earth itself, and the Earth in its orbit around the Sun. (You can select any combination of these, to avoid overloading the screen at the start.) You can select particular points on the Earth's surface and watch them move in and out of daylight at different times of day, so that you can see how day length changes. You can alter the angle at which you are viewing the Earth or its orbit. At the same time, you can superimpose graphs to show how temperature or light intensity varies. These graphs develop as the Earth turns.

Lessons from *Earth and Sun*
This software illustrates the point that multimedia can change the speed at which things happen; a day or a year can be compressed into a few seconds.

You can select points of interest on the Earth; for example, if you have students who have family members in Pakistan or Australia, select one of these, plus your home town. In which place does the Sun rise first? Why should you not telephone relatives in Australia when it's tea-time in the UK? Why are Pakistan's seasons different from the UK's? Personalising software can greatly increase its relevance and its impact.

Software like *Earth and Sun* can seem daunting at first, so it is provided with ready-made 'slideshows'; these are sequences of screens in which the different controls have been set so that you can page through the slideshow to make a coherent lesson. Students can work through the slideshow at their own pace, following instructions on an accompanying worksheet. 'Get to know' slideshows are also provided, to give teachers a lesson in 'driving' the software. Once you are familiar with the way it works, you may decide to make your own slideshow, so that you and your colleagues can reproduce a particular sequence of screens.

Moving molecules
We can't see atoms and molecules, but today's students have the great advantage that they can see good representations of them using multimedia. Particles can be shown packing together to make solids and liquids, and moving freely as gases. They can be seen diffusing, colliding and reacting.

Sunflower Learning's *Diffusion* can help to get across the basic mechanism of diffusion. How does the random motion of myriads of particles result in the spreading out of a liquid or gas? Students can see differently coloured particles moving randomly, colliding with one another, and gradually filling a space (Figure 3.1, overleaf).

Figure 3.1
A screen from Sunflower Learning's Diffusion.

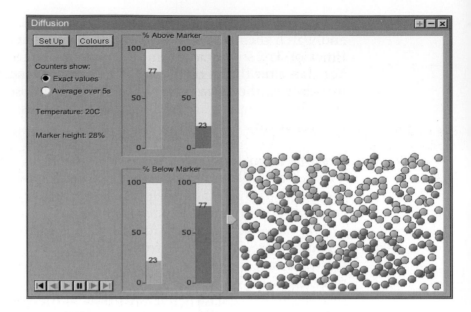

Lessons from *Diffusion*

Chemistry benefits from the possibility of making visible the invisible. Much of what students learn from software like this is implicit. For example, they learn that particles vibrate more or move faster at higher temperatures. They also learn that particles do not appear or disappear in a chemical reaction but merely rearrange themselves. Once students have grasped this, the idea of chemical equations can seem more natural.

Diffusion comes with a range of on-screen worksheets. The controls have been appropriately set, so there is little danger of the user making inappropriate selections or wandering off the point. Students are presented with a sequence of questions. They fill in answers on-screen, and print off the final completed worksheet.

Photosynthesis

Photosynthesis is a topic that is generally covered in the later years of secondary science education. A useful lesson can be built around a tool called *Photosynthesis*, which is part of PLATO Learning's *Multimedia Science School*.

This tool has two parts, called 'scenarios'. In the first scenario, students investigate the factors that determine the rate of photosynthesis in a specimen of pondweed. This is a standard school experiment, and students may well have carried it out for themselves in the lab.

Using this software package, students have the opportunity to alter four factors: the temperature, the carbon dioxide concentration, the light intensity and the light colour. The controls are simple sliders or buttons (see Figure 3.2); by adjusting these controls, they can find out about how each one affects the rate of production of oxygen by the pondweed.

Figure 3.2
The first scenario from PLATO Learning's Photosynthesis.

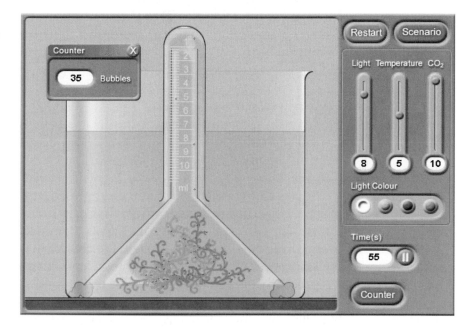

Figure 3.3
The second scenario from PLATO Learning's Photosynthesis.

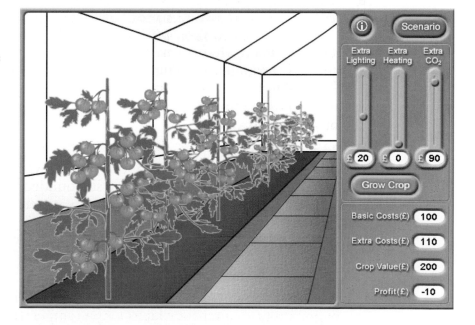

In the second scenario (Figure 3.3), students are put in the role of tomato-grower. They have to manage a commercial greenhouse in such a way that the crop they produce is profitable. To do this, they must use what they have learned in the first scenario. They can operate controls to change the light intensity, the temperature and the carbon dioxide concentration inside the virtual greenhouse. They see the tomato plants grow and bear fruit, and their costs are compared with the crop value, so that they can see whether or not they have made a profit.

Lessons from *Photosynthesis*

The most important lesson from this example is that the second scenario provides a challenge. Without expenditure on lighting, heating and carbon dioxide, the crop makes a loss. Can you grow your tomatoes at a profit? If you know which factor has the strongest influence on photosynthesis, you can make a good guess at how to manage the greenhouse. Who can be the first to make a profit? Who can make the most profit? Students find a challenge like this very motivating, in the same way that computer games are motivating.

Another point to note is that, in the second scenario, students make use of what they have learned in the first scenario. They use what they have learned about photosynthesis, but they also use their understanding of the multimedia controls, which are very similar in the two scenarios.

As with other tools in *Multimedia Science School*, this one comes with a ready-made slideshow and worksheet to guide students through the activity in a structured way. However, the ideas behind it are mostly self-evident, and many students would grasp the procedure from a simple, one-minute introduction from you.

Radioactivity

Older students need to learn about radioactive materials and the radiation they produce. We are surrounded by radioactive substances and their radiation constantly penetrates our bodies. People are fearful of this invisible radiation, but they cannot make realistic judgements unless they have a scientific understanding of its effects. Should I accept radiotherapy treatment? Is nuclear waste stored safely?

Newbyte's *Radioactivity – Penetration* is one of a set of multimedia programs on radioactivity. Students can perform virtual experiments using different sources, and with different materials and thicknesses between the source and the Geiger counter. They can learn to distinguish alpha, beta and gamma radiations by their penetrating powers, and identify the radiation from different sources.

Lessons from *Radioactivity – Penetration*
This program is best used alongside real experiments. Students need to see real radioactive materials, and learn how they can be handled safely. They need to watch experiments using Geiger counters. However, they cannot carry out such experiments themselves, because radioactive sources must not be handled by students under the age of 16. So this software will allow students to perform experiments for themselves in a safe way, once they have seen the real thing demonstrated.

An additional value in multimedia like this is that it shows the invisible. We cannot see alpha, beta and gamma radiations, but they can be represented on the screen. In addition, background radiation can also be represented. This can strongly reinforce ideas that may have been presented during a teacher exposition, or perhaps on video.

◆ *Conclusion*

To conclude this chapter it will be useful to consider some of the common debates about the benefits of multimedia in the classroom.

Are we trying to teach science or ICT?
The answer is science, of course, but necessarily aspects of ICT also. We want our students to be able to use new multimedia resources with confidence, in the same way that they can use new books or new laboratory equipment. This book describes the many ICT skills that are valuable in science. Interaction with multimedia will teach new skills, or reinforce existing ones.

As we have mentioned, there is always a danger that the ICT itself becomes the focus of the lesson, rather than the ideas you want to get across. So you need to decide how to build multimedia into your lessons so that students see it as a useful tool rather than as an end in itself.

Laboratory experiment or multimedia simulation?

Multimedia should be an enhancement, not a replacement. You may have wondered, "Isn't it better to do an experiment than to watch a simulation?". Yes, it often is. But a simulation has the following advantages in some situations.

- It can allow you to repeat an experiment several times, quickly.
- It can allow your students to do a dangerous or expensive experiment, perhaps after a real demonstration.
- It can help to focus students' attention on the outcomes, rather than the method.
- It can extend the experiment, by allowing you to change more variables.
- It can draw graphs automatically.
- It can show up what is invisible in an experiment. (How are the molecules moving?)
- It will appeal to students who have more visual learning preferences.
- It gives students another route to understanding.

In a similar way, looking down a microscope at a student-prepared slide may be an ideal. Looking at a micrograph of the same tissue in a book may give a clearer view. A textbook diagram can perhaps show the essential features most clearly. But all three, working together, can get the concept across best. Multimedia is the same – but it moves! Experimentation, textbooks and multimedia, combined with the guidance of a teacher or lecturer, can bring the best of all worlds.

CD-ROM or www?

Multimedia is often more convenient on a CD-ROM. You don't have to worry about Internet access or the speed of your connection. But software can go out-of-date (you may have an old version) and you generally have to pay for CD-ROMs, whereas many Internet sites give you free access and you may be able to download software onto a desktop or laptop computer.

You will find collections of educational Java applets on the Web. (An *applet* is a small application; *Java* is the language in which they are written.) An interesting collection of physics applets, for example, is at www.phy.ntnu.edu.tw/ntnujava/. An applet is a pared-down form of multimedia; it should require little supporting information, and its controls should speak for themselves. Some may be rather quirky and individualistic, but in general a good applet can be very useful for presenting a simple idea in a clear and unambiguous way.

Why do I pay so much for multimedia?
Good multimedia is produced by a team of experts: programmers and graphic designers do the bulk of the work, alongside educational and scientific advisers. Materials may have been trialled in schools and revised in the light of experience. Each of the products described in sections 3.2 and 3.3 above represents hundreds or even thousands of hours of work. That's expensive.

Much good software, however, is available free. It may have been produced with commercial sponsorship, or by an enthusiast working alone. It can be very good – try to assess it against the ideas presented in this chapter.

◆ *Multimedia producers*

Crocodile Clips Ltd
Software for physics and chemistry.
www.crocodile-clips.com
Dorling Kindersley
Educational software products, available in the UK from Global Software Publishing (GSP) Ltd.
www.gsp.cc
Newbyte Educational Software
A range of software originated in Australia.
www.newbyte.com/uk
PLATO Learning (formerly New Media)
Multimedia Science School (an ever-expanding resource for ages 11–16) and many other materials for science education.
www.platolearning.co.uk
Sunflower Learning
A range of tools under the title of *Multimedia Library for Science*.
www.sunflowerlearning.com

Making students creative with ICT

Tony Sherborne

♦ *Introduction*

Getting students to tell stories is a powerful way of helping them to learn. Every teacher knows how describing ideas to others enables you to improve your own understanding. Stringing the ideas together into a narrative makes them more memorable, both to the audience and the storyteller.

Science is full of abstract and complex ideas, from the particle model of matter to the nature of scientific evidence. Explaining these purely in words can just create a conceptual fog, obscuring the important ideas from the students. You need clear images to penetrate the mist, which is why we use illustrated textbooks and computer animations. But as good as these are, they suffer from a disadvantage. The images in diagrams and in computer models are highly stylised: they are not always easy for students to relate to reality.

This chapter offers an alternative: for students to create visual stories based on the model that is in their heads. This approach takes a little more time, but in my experience almost guarantees that afterwards they can provide a clear explanation, from memory, of the scientific concept. It can also be a lot of fun.

Here we will outline two strategies for getting students to tell stories:

♦ through 'animation'
♦ through 'video editing'.

Both essentially involve making decisions about what images to put on the screen, and selecting the order in which to present them. It is this decision-making process which actively engages students with the material, and clarifies the conceptual model in their heads.

Difficult as these techniques may sound, you will see that it is possible to make a convincing attempt without using anything more complex than Microsoft PowerPoint. To make the techniques easy to adopt and apply, they are presented here in the form of actual case studies.

4.1 Animating with PowerPoint

People misunderstand PowerPoint. They use it to present slide after slide of words. That is not what it is good at: important text would be easier for students to remember as bullet points on a handout. Presentation software should be used with a different mindset: to help the audience visualise the ideas (complementing the words spoken by the presenter).

PowerPoint has sufficient animation features to visualise scientific processes in action. For instance, in chemistry you can easily 'animate' a chemical reaction, by making the reactants (graphics) come onto the screen, move together and then 'morph' into a new substance. PowerPoint software is easy to use and available on almost any school computer, unlike the more powerful purpose-designed animation software.

The case study: acid rain

We will now look at a published animation activity for teaching environmental chemistry. It is called 'Sulphur dioxide in the atmosphere' and is on the CD-ROM *Seeing Science with CCLRC*. Students create animations to explain (and therefore learn) the processes by which acid rain is produced, and the effects it has:

1. How sulphur dioxide is produced in power stations and is transformed in the atmosphere into acid rain.
2. The effect of acid rain

   - in the countryside: on fish and trees
   - in the city: on people and buildings
   - in the atmosphere: as a greenhouse gas.

Students need objects (graphics) to animate, from background landscapes to molecules or movement arrows. The resource provides a large selection (see Figure 4.1), ready-to-go inside a PowerPoint template (for version 2000 or later).

Since animation takes time, students (working in pairs) are asked to tell one part of the story each (the story is divided into four acts, with two scenes in each). At the end, they can watch the others to see the complete story. To help students move from a 'bullet-point' to an 'animation' mindset, the materials adopt a theatre metaphor. Each animation is a 'scene', which takes place in a 'setting', and the graphical objects are described as 'actors'. Each pair of students is given an 'animation brief' like the ones in Box 1 on page 77.

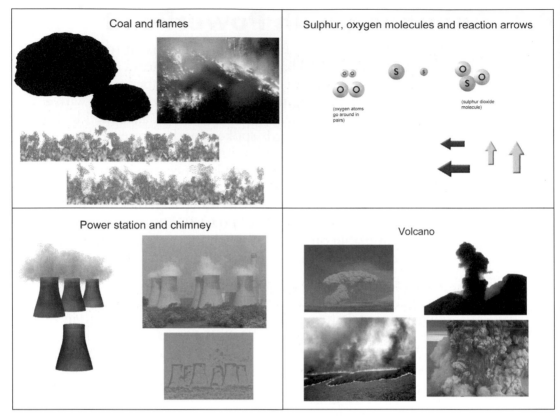

Figure 4.1
Examples of graphics or 'actors' for animating, from Seeing Science with CCLRC.

Table 4.1 *PowerPoint animation features.*

What it can do	Example of its use in this lesson
Move actors from offstage into the middle of the screen	Make molecules of sulphur and oxygen come together for a reaction
Make actors appear or disappear (at a particular time/in a particular order)	Construct a pH meter scale to show acidity in a lake increasing over time, with coloured bars representing pH that successively appear above each other
Transfer to a duplicate slide which has been altered slightly (on a click or automatically)	Illustrate step-by-step the process of sunlight hitting the Earth, being reflected as heat, and absorbed by greenhouse gases
Move actors along a defined path (only PowerPoint 2002 onwards)	Show nutrient particles in the soil being drained out and running into a lake

---

Box 1. Example animation briefs

Act 1 Scene 1. *How sulphur dioxide is created*
Fossil fuels like coal contain sulphur as an impurity. Coal is burned in a power station. The hot air (oxygen) also reacts with the sulphur. It forms the acid gas sulphur dioxide. This escapes from the chimney. Sulphur dioxide is also released from erupting volcanoes.

'Setting':   Inside the power station.
'Actors':   Coal and flames, sulphur, oxygen, sulphur dioxide, chemical reaction 'arrows', power station, chimney, volcano, meters.

Act 4 Scene 1. *Effects of sulphur dioxide on planet Earth*
Sunlight hits the Earth. It is reflected off the surface as heat waves. These would normally escape. But there are 'greenhouse gases' in the air, like sulphur dioxide. These gases absorb the energy. The trapped heat warms the atmosphere, like a giant greenhouse.

'Setting':   Earth's atmosphere.
'Actors':   Sun, Earth's surface, sunlight and heat arrows, meters.

The activity is divided into two stages:

1. Storyboarding: planning the animation as a series of sketches showing the sequence of 'action'.
2. Animating: using PowerPoint to implement the storyboard – putting 'actors' onto slides, and moving them around to tell the story.

The activity gives scope for students to be imaginative in their choice of elements and how they move, thus making it more memorable. It is also differentiated in how students respond. Some will attempt to describe only the surface features of their story, while others will try to show chemical reactions in terms of particle rearrangements. The teacher can use the complexity of the animations as a measure of students' understanding.

Storyboarding

Students will develop a better understanding if they think through the details of the process and how to visualise it, before getting into the detail of the software. At this stage they need a clear idea of the kind of animation they are aiming to produce. Among the PowerPoint templates are finished examples that illustrate the range of simple animation features available. As Table 4.1 (page 76) shows, with a little imagination, these features can do most things you need to tell the story.

The animation brief sets out the main action in the scene. The challenge for students is to interpret these directions. Having to 'sketch' the scene forces them to visualise what they want the actors to do. For Act 1 Scene 1 (see Box 1, page 77), they might break down the action into steps like these:

Figure 4.2
Example of a student storyboard for Act 1 Scene 1 of the brief.

Picture 1	Picture 2	Picture 3	Picture 4
Power station and chimney, against sky background.	Zoom in on the coal above a flickering flame showing yellow bits – the sulphur impurity.	Oxygen particles appear and move close to the sulphur.	Sulphur has combined with the oxygen to make sulphur dioxide, which moves up the chimney.

Animating

With a storyboard, the actual job of animating becomes a lot easier. Now students can use PowerPoint to create slides, placing actors into position, and then play with the 'custom animation' feature to move them about, to control the order in which actions happen, or to make the timing automatic.

In addition to the provided graphics, students can add their own lines or shapes using the 'draw' toolbar, or import graphics from the Web. Though the emphasis is on telling the story in pictures, students can also be encouraged to add text labels where necessary to communicate the ideas more clearly.

Finally, students present their animation. They can add a spoken commentary, providing background information to support the visuals. This could even be recorded into the

PowerPoint file to keep as a permanent record. Students in the audience can be involved in assessing how well the ideas in the story were communicated, for instance by checking how much they can remember after the presentation.

This animation technique can be adapted to many other topics in science. To make it work only requires putting together PowerPoint templates of relevant graphics. In most cases you can source these from clip-art CD-ROMs or the Web (try Google™ Image Search).

4.2 Video editing

Digital video can 'motivate and engage a wider range of pupils than traditional teaching methods, so providing greater access to the curriculum' (*Using Digital Video in Teaching and Learning*, Becta 2002). If the potential of video as a learning vehicle is great, so are the challenges it poses. How do you justify the time needed for students to make video in an overcrowded curriculum? How do you manage with perhaps one or two video cameras for a whole class? And how do you control such an open-ended task to ensure students achieve the intended learning outcomes?

The following case study explores an approach to using digital video which preserves a lot of the motivation, but minimises the requirements of time and equipment; students focus on the editing process. Editing is the process of turning the already filmed material into the final programme, by selecting, ordering and cutting down the clips to size.

The Becta report concluded that making video helps students develop a wide range of key skills. The editing part of the process is a structured decision-making activity, and is arguably as effective at developing thinking skills:

- information processing – sorting and sequencing clips
- reasoning – giving reasons for opinions in writing commentary
- evaluation – judging the value of arguments in deciding which clips to include
- communication skills – how to tell a story visually (and it promotes collaborative discussion, if students work in pairs)
- problem solving and creativity – the same material can be edited together in different ways.

However, with all these advantages, video editing is only possible if you can provide students with the already filmed raw material in the first place. For this case study, the video clips were specially produced. To apply the technique to a different topic, you could either take video clips from a CD-ROM or DVD (e.g. one of Channel 4's) or you could brief a group of students to film the required material outside lesson time.

Video editing requires software, which some schools now have access to. For PCs, you can use the free Windows Movie Maker (if you have Windows ME or XP), or you can buy a product such as Pinnacle *Studio* (currently about £50 a copy). For Macs there is the popular iMovie, also free. However, even if you do not have proper video editing software, you can use trusty Microsoft PowerPoint to simulate the process, as described in the case study.

The case study: ideas and evidence

We will now look at a video editing lesson created to teach 11–14 students the nature of scientific evidence. The 'Ideas and evidence' section of the National Curriculum for Science requires students to study a range of contemporary issues in order to understand three big ideas in science: evidence, experimentation and explanation. The lesson is part of the Science Consortium's *Online CPD for Science* package (www.cpd4science.co.uk).

The issue chosen for this lesson was: "Does homeopathy work?". This is an interesting and controversial topic, ideal for making into a video. Interviews were scripted and filmed with three people who have different views about homeopathy: a medical researcher, a doctor and a homeopath.

The video material provided for students covers the main concepts of ideas and evidence:

1. *The importance of evidence to back up claims.* Millions of people use homeopathy and believe it is an effective treatment. However, is there any scientific evidence that it works? Apparently, for some minor conditions such as allergies, there may be.
2. *Evidence comes from carrying out good experiments.* Many experiments have been conducted into homeopathy. Famously, a French scientist made a sensational discovery that molecules of water have a memory. However, his findings were completely overturned when the experiments were repeated in more controlled conditions.

3. *Scientists create theories to help explain phenomena.* Most scientists are sceptical of homeopathy because it contradicts the well established particle model of matter. Homeopathic solutions are made by diluting chemicals in water so much that in all probability there are no molecules of the active substance left. The 'scientific' explanation for any health benefits of homeopathy is that it just works by the placebo effect: if you believe in it, then your mind can heal the body.

The lesson begins by introducing the context. Then students are set the task with a 'programme brief', like the one in Box 2 below.

Box 2. Example programme brief

Make a 2-minute section of the programme to answer *one* of these questions:

Section 1
Patients believe homeopathy works. But is there scientific evidence that it can cure people? (*use Q1 interviews*)

Section 3
Many scientists think that homeopathic remedies do not work. What is the scientists' explanation for why patients get better? (*use Q3 interviews*)

Use software to edit together video clips of three people with different views. Write a commentary to introduce and sum up the debate, and introduce the speakers.

The process of editing the video is structured into three parts:

1. Selecting clips: choosing the contributors' answers most relevant to the brief
2. Assembling the movie: deciding how to tell the story, by ordering the clips
3. Writing commentary: introducing and summarising the contributors' arguments.

Selecting clips
The 2-minute requirement of the example brief is a key factor in the learning process. Students are given about 6 minutes of interview material for each section. They have to boil it down

to the interview answers that are most relevant to the question set in the brief. Students need to play the clips through and carefully assess how well each contributor's answers throw light on the concepts of evidence, experimentation or explanation. A table in which to note down the key points from each answer makes this easier.

All the video editing packages mentioned earlier have similar features to make this process quick and easy:

- you 'import' all the individual clips into a 'project'
- you can drag the clips you want to use onto a 'timeline'.

If instead you want to use PowerPoint for the editing, then:

- create a new slide for each video clip
- 'insert movie' to import the video clip
- play the clips by running the slideshow, and clicking the slides.

Assembling the movie

Having chosen the most important clips, students now have to decide which order to put them in. The thinking involved can be challenging for students. They have to construct a clear, logical argument from a series of individual clips. In the published lesson, students are encouraged to structure their thinking by using a storyboard. See, for example, Table 4.2.

Table 4.2 *Planning the order of clips.*

Part of the template storyboard		Which clip?	Approx. time
First commentary/ introduction	Words for commentary: (people speak at 3 words per second)	e.g. 'poster and sign'	
First interview	Name: (e.g. Doctor Mark Loughlin)	e.g. 'doctor 3'	

The editing software makes the mechanics of assembling the movie easy. You can:

- re-order clips on the timeline, by dragging and dropping
- play through the whole movie, to check whether it makes sense and keeps to time
- 'trim' each clip, removing the parts that are less relevant.

Figure 4.3 *Video editing with a) Windows Movie Maker and b) PowerPoint.*

a)

b)

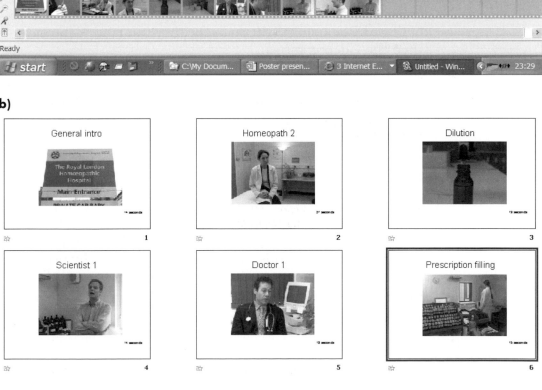

PowerPoint can simulate this part of the editing process:

- use the 'slide sorter' to put the selected clips in the order you want
- play the clips by running the slideshow
- manually play through the movie, by clicking to the next slide after each video clip finishes playing.

Writing commentary

With any documentary, part of the storytelling is in the commentary. For this video, students can write several short pieces, of about 15–20 seconds in length (60 words):

- an introduction to make viewers interested in the debate and to introduce the first contributor
- 'cutaway commentaries' to help the viewer make sense of the interview clip that has just been seen and to introduce the next (some general 'filler shots' are provided for this)
- a conclusion which sums up the arguments and answers the question in the brief.

It is a good idea for students to plan the commentary on the storyboard. If desired, they can then record this commentary with a microphone directly into the video editing software or the PowerPoint slide. Alternatively, they can read it through when the video is shown.

Playing the finished videos provides a climax to the activity. The work can be assessed in terms of how well the chosen clips, the order of presentation and the commentary all help to communicate the nature of evidence, experimentation or scientific explanation.

◆ *Conclusion*

This chapter has shown two highly motivating ways for students to learn, both by telling visual stories through the power of simple software. These approaches take a little longer than traditional methods, but they make up for it by providing a rich learning experience that helps the development of higher-order thinking skills.

◆ *Reference*

Becta (2002) *Using Digital Video in Teaching and Learning*
www.becta.org.uk/corporate/publications/documents/
Using%20Digital%20Video.pdf

◆ *Resources*

Online CPD for Science for video editing and many other ICT
applications in the classroom.
www.cpd4science.co.uk

Pinnacle *Studio* for video editing.
www.pinnaclesys.com

Seeing Science with CCLRC for animating with PowerPoint.
www.seeingscience.cclrc.ac.uk
CD-ROM available free via the website or from:
CCLRC Rutherford Appleton Laboratory, Chilton, Didcot,
Oxon OX11 0QX. Tel: 01235 445950

Further animation and video editing materials are available
from the *Science UPD8* website:
www.upd8.org.uk

Mathematical methods using ICT

Peter Myers

◆ *Introduction*

Perhaps the area of the science curriculum that lends itself most to the use of ICT is the treatment of numerical data. In this chapter we explore several different ways in which ICT could be used to deal with numerical data, ranging from simple graph plotting for experimental data to the use of multimedia simulations to model complex ideas. For each example given there is a description of how it works, what the benefits are and how the methods could be used elsewhere.

The use of technology enables learners to progress at their own rate, freeing up teachers to assist weaker students or extend the more able. The technology also concentrates the students' minds on the activity itself rather than becoming bogged down in the mechanics of processing data. I hope you find that trying some of the activities or similar ones will demonstrate their effectiveness.

5.1 Plotting experimental data

One of the most obvious uses of ICT is for plotting graphs of experimental data. There are a number of reasons why we may wish to encourage our students to use ICT to do this. Students find graph drawing tedious, and the use of ICT can help them to focus their attention on the analysis of the graph rather than the mechanics of plotting it. In addition, the final product can be an attractive object; many students value this aspect of the presentation of their work.

An example: conducting solutions

Here is an example using Microsoft Excel. The data is from a simple experiment in which the electric current through a salt solution is measured as the concentration is increased.

Excel can be bewildering for new users when it is first launched, but the key is to focus on the areas of interest and not on the plethora of functions. For this exercise we are interested in entering some data and plotting a graph so those are the areas to consider first.

The initial grid needs to be filled with data and it is typically more convenient to use the format where the data is in columns. Figure 5.1 shows a simple spreadsheet with just two columns of data. To start plotting an appropriate graph we need to first select the data, and then click the chart wizard button on the toolbar. The dialogue box in Figure 5.2 then appears. This takes you through a four-step procedure, resulting in a graph incorporated in the spreadsheet.

Step 1
The graph-plotting options are quite comprehensive but for most experimental results the scatter graph option is best. Having selected the graph type it is then possible to select a sub-type; since in this example we don't want to see the points joined up, it is better to use the 'points only' sub-type (top left).

Step 2
Choosing 'Next' shows a preview of the graph you will produce. The data we used were entered in columns so no changes are needed at this point, although it is possible to correct a graph if data have been entered in rows rather than columns.

Figure 5.1 *The data in an Excel spreadsheet.*

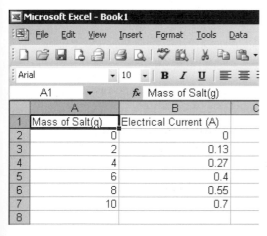

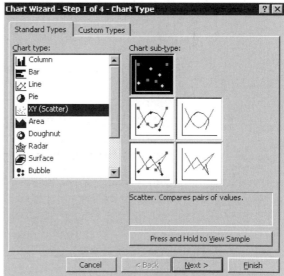

Figure 5.2 *The chart wizard dialogue box.*

Step 3

Again choose 'Next', so that you can label the axes and title the graph. The preview automatically updates the graph to reflect the changes. The tabs at the top of the dialogue box allow you to change the format of the graph now, or you can leave the default settings. With Excel it is possible to amend anything later should you wish to.

Step 4

This determines the placement of the graph you have produced. Select the *As object in* option to place the graph with the data so the two can be compared – see Figure 5.3. (The *As new sheet* option places the graph on a different sheet from the table.) The graph can be formatted to suit individual taste by selecting the graph and double clicking the part you wish to change; e.g. to change the *x*-axis, double click on it and make selections from the dialogue box that appears.

Figure 5.3

The graph together with the data.

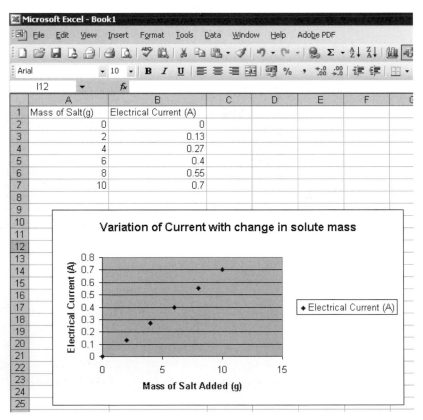

Why it's worthwhile

A frequently discussed question is whether Excel (or another software package) should be used by secondary school students to draw graphs. In its favour, Excel allows the student to quickly produce the graph and then concentrate on the interpretation of the graph and the analysis process. It can also give students who have difficulty in determining appropriate axis scales the opportunity to practise interpreting graphs without becoming bogged down in the mathematical process of fitting the axes to the paper.

Against this, it is suggested that using a computer-generated graph removes the student from the first-hand appreciation of the data, and that heavy reliance on computers means the student may be unable to spot, for example, erroneous data.

I believe there is a compromise that lies somewhere between these two views. Use of graph-plotting software can ensure that students get accurate graphs of their data, but at the same time it is important to encourage students to identify anomalous points, to draw a line of best fit for themselves and to format the graph to fit the data. They are then getting useful practice in many aspects of graph plotting while not being restricted by a lack of mathematical skills.

Using the example elsewhere

An extension of this example that many older secondary students will meet is based on measuring the resistance of different lengths of wire. Figure 5.4 overleaf shows how Excel can be used to produce complex tables as well as graphs. This illustrates how repeat readings can be handled.

The first column (length of wire (m)) has merged cells for each value. Resistance values have then been calculated for each of three experimental runs using $R = V/I$; calculated average values are shown in merged cells in the final column (average resistance (Ω)). To select the first and last columns for graph plotting, select the first column and then hold down the *ctrl* key as you select the final column. This allows the selection of two non-contiguous sets of data for a single graph.

The graph in the example has been plotted without a line of best fit, to allow the line to be drawn in manually.

Figure 5.4 *A complex table and associated graph produced in Excel.*

Factors affecting the Resistance of a Wire - Length

Length of Wire (m)		Run 1	Run 2	Run 3	Average Resistance (Ω)
1	Measured Potential Difference (V)	11.81	11.79	11.77	
	Measured Current (A)	3.50	3.60	3.60	3.31
	Calculated Resistance (Ω)	3.37	3.28	3.27	
2	Measured Potential Difference (V)	12.01	12.11	11.97	
	Measured Current (A)	2.00	1.80	2.10	6.14
	Calculated Resistance (Ω)	6.01	6.73	5.70	
3	Measured Potential Difference (V)	11.90	11.99	12.20	
	Measured Current (A)	1.20	1.10	0.80	12.02
	Calculated Resistance (Ω)	9.92	10.90	15.25	
4	Measured Potential Difference (V)	12.10	12.01	12.02	
	Measured Current (A)	0.80	0.90	0.90	13.94
	Calculated Resistance (Ω)	15.13	13.34	13.36	
5	Measured Potential Difference (V)	11.75	11.99	12.11	
	Measured Current (A)	0.60	0.60	0.70	18.96
	Calculated Resistance (Ω)	19.58	19.98	17.30	

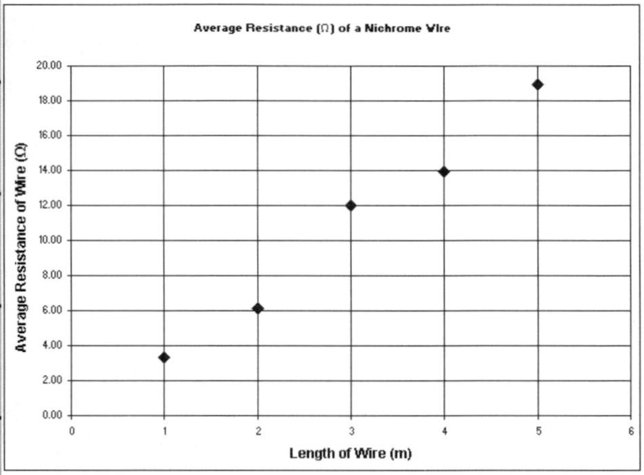

5.2 Developing ideas

The principle of moments

This example, and the one that follows in section 5.3, are among the relatively few opportunities to use ICT-based calculations with younger secondary students. It looks at the principle of moments, often referred to as the see-saw rule. A mouse can balance an elephant if the see-saw is long enough (Figure 5.5).

The object of this exercise is to provide a framework for students to develop the idea for themselves. It could be carried out in several ways, e.g. by using experimental data and using the spreadsheet to process it, or by giving the students some partial data and allowing them to use the mathematical capabilities of Excel to determine the rule. Here we are going to use experimental data.

Figure 5.5
Balancing an elephant with a mouse.

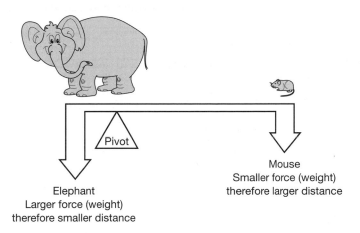

The experiment involves balancing the see-saw by applying different masses to either side of it, at different distances from the pivot. Students use 1 g and 10 g masses (equivalent to 0.01 N and 0.1 N), and the see-saw is 30 cm long so that the maximum distance from the pivot is 15 cm. This assumes that the see-saw will balance on its own.

Figure 5.6 overleaf shows an Excel spreadsheet in which students enter their values for the distance from the pivot. Values of weight are provided. The spreadsheet is locked (see below) so that the only cells that can be edited are the ones in which students enter their data.

In Figure 5.6 the whole spreadsheet is displayed, including the normally hidden formulae. The formulae can be turned on and off using the 'options' dialogue under the Tools menu. The cells that the students will edit are shaded in grey; these are unlocked using the Format Cells command. Columns C and G are hidden in the version presented to the students. The cells on the right ('Checking the Data') can be both locked and hidden using the same commands. The formulae in columns C, G and J are used to check the students' data. The final column will indicate whether a given set of data will result in a balanced see-saw.

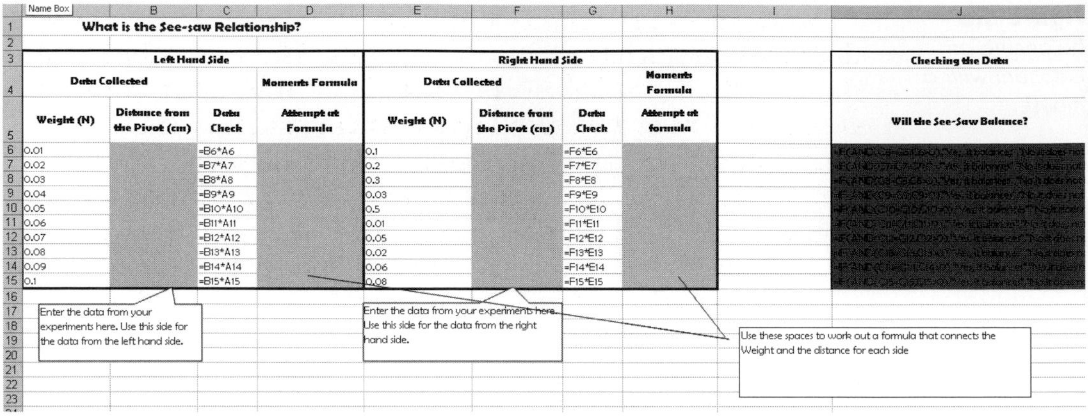

Figure 5.6
The full Excel spreadsheet for the moments experiment.

Before presenting the students with this spreadsheet it must be locked. Turn on the Protection toolbar and click the Protect Sheet icon. The dialogue box allows you to specify what, if any, changes the students can make and typically all of these need to be unchecked. Figure 5.7 shows the spreadsheet as presented to the students. The 'Checking the Data' column is not essential but is there to discourage the students from guessing values and filling in the table randomly.

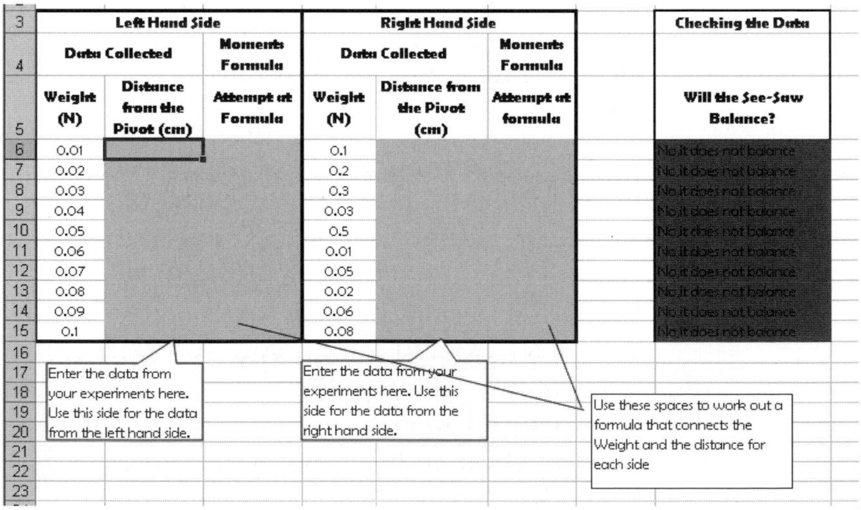

Figure 5.7
The student version of the spreadsheet shown in Figure 5.6.

Students can now carry out their experiment and find the distances at which the see-saw balances. The spreadsheet will then tell them if it balances or not. The 'Attempt at Formula' columns provide a working space where students can try out simple calculations to determine a relationship between the data. They already know that there is a relationship that works, so it is up to them to try combinations of add, subtract, multiply and divide to determine a relationship. Depending on the ability of the students you may wish or even need to give extra help and support.

Why it's worthwhile

Using a spreadsheet like this allows students to try calculations very quickly and (once they have got the idea) with minimal teacher input. The hard work comes in persuading the students that this is within their capability. It is possible for most students to come up with sets of data that both work and are simple enough for them to understand; alternatively, you can insert more challenging data for the more able.

Using the example elsewhere

The exercise can easily be adapted to other experiments where there is a simple mathematical relationship in the numerical data generated, for example the derivation of Ohm's law ($V = IR$), Newton's second law (force = mass × acceleration), and transformer rules. The spreadsheet acts as a place for students to record their data and builds in a check to make sure that they do not cheat or enter incorrect data.

5.3 Handling large data sets

One of the powerful uses of Excel and other spreadsheet packages is in handling large quantities of data. The activity below is based on data from *Census At School* (see page 101).

How much have you grown today?

The aim of this exercise is to estimate average change in height between year groups. Records of height and school year for thousands of young people are available. In this example, records for 258 students were downloaded into a spreadsheet. These were sorted into order by year (using Data – Sort) and then split into year groups. The average heights were calculated using the AVERAGE function and the differences between year groups calculated very quickly. All this data can be processed in a matter of minutes and by individual students. (You might consider asking your students to start processing this data by hand, simply to reinforce the power of ICT.) The first two rows of Table 5.1 show a summary of the results.

The data (which of course is very real) can now be investigated. First of all the difference between years can be used to determine the rate of growth for a particular year group – see row 3. The table can now be completed by adding formulae to calculate the growth in different time periods.

Students can now be asked to comment on the results. For example, could we determine the hourly growth of a student in the class? How accurate are the results? Does each year group grow at the same rate and, if not, why not? Can you think of other factors apart from age that will affect growth? How could you go about investigating them?

Table 5.1 *Completed table of growth in different time periods, calculated by Excel.*

Year	3	4	5	6	7	8	9	10	11
Average height (cm)	128.65	128.74	137.10	141.39	149.45	156.32	161.39	171.00	170.45
Difference between years (cm)		0.09	8.35	4.29	8.06	6.87	5.06	9.61	–0.55
Growth per month (cm)		0.007	0.696	0.358	0.672	0.573	0.422	0.801	–0.045
Growth per week (cm)		0.002	0.161	0.083	0.155	0.132	0.097	0.185	–0.010
Growth per day (cm)		0.0002	0.0229	0.0118	0.0221	0.0188	0.0139	0.0263	–0.0015
Growth per hour (cm)		0.0000	0.0010	0.0005	0.0009	0.0008	0.0006	0.0011	–0.0001

Why it's worthwhile

The use of large quantities of data is one of the primary ways scientists make deductions and reduce errors. The use of this amount of data makes the results much better than if we sampled a single class. The example here only used a small subset of the available data, but the data can be tailored to a regional group and could include factors that might be related to a particular measurement. It is even possible to submit your own data to the archive, and so give students the sense of contributing to a larger pool of data.

Using the example elsewhere

Any population experiment in biology needs a large body of data in order to become representative, and so it becomes much more manageable using ICT. Datalogging experiments typically generate large quantities of data. When the results are examined in Excel it is found that there is too much data to deal with by hand and so our only recourse is to ICT.

5.4 A complex model

Stopping distances

Older secondary students can use ICT to look at the factors affecting the stopping distance of a moving vehicle. This lends itself to a graphical demonstration in Excel. The following extract from the AQA Double Award Modular Science Specification indicates the material we are trying to cover.

The stopping distance of a vehicle depends on:

- the distance the vehicle travels during the driver's reaction time;
- the distance the vehicle travels under the braking force.

The overall stopping distance is greater if:

- the vehicle is initially travelling faster;
- the driver's reactions are slower (due to tiredness, drugs, alcohol);
- there are adverse weather conditions (wet/icy roads, poor visibility);
- the vehicle is poorly maintained (e.g. worn brakes/tyres).

These two key statements give us an opportunity to develop a model to allow students to change conditions and see the effect they have on thinking, braking and stopping distances.

The example here illustrates the use of two tools from the Forms toolbar, the Spinner and the Combo Box. Students are presented with a spreadsheet as shown in Figure 5.8. The Spinner is on the left, and allows the user to change the car's speed in steps of 10mph. There are three Combo Boxes, used for selecting particular conditions. Making any selection from the many possible combinations results in a change to the bar chart.

Figure 5.8 *The Excel spreadsheet for stopping distance, as seen by the student.*

	A	B	C	D
1				
2	Speed of the car (mph)	Conditions	Driver	Car
3		Dry	Alert	Good Condition
4		Wet	Drunk or tired	Poor Brakes
5				Worn Tires
6				Poor Brakes & Worn Tires
7	▲	Dry	Alert	Good Condition
8	▼			
9	20	1	1	1
10		=IF(B9=2,1.5,1)	=IF(C9=2,2,1)	=IF(D9=2,1.25,IF(D9=3,1.5,IF(D9=4,1.75,1)))
11				
12		Thinking Distance	Braking Distance	Stopping Distance
13	Dry Conditions	=A9/10*3	=A9^2/65	=B13+C13
14	Chosen Conditions	=B13*C10	=C13*B10*D10	=B14+C14
15				

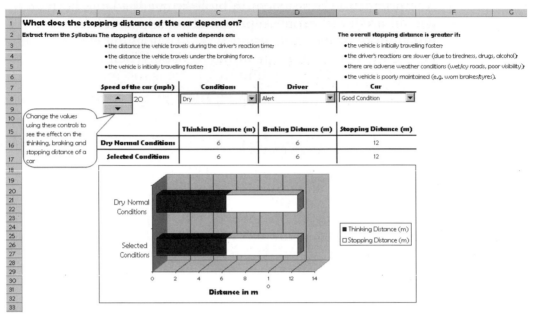

Figure 5.9 *The hidden spreadsheet, showing the formulae used to calculate the thinking, braking and stopping distances.*

Behind the user-friendly version of Figure 5.8 lies a hidden spreadsheet, as shown in Figure 5.9, which includes the appropriate formulae for calculating the outcome of any choice of parameters.

Why it's worthwhile

The statements in the specification are reasonably obvious and once presented with a few graphs of the data they become tedious for most students. Presenting the exercise in this way requires more initial setting up but then it becomes more interesting for the students. The ability to change and compare the values in real time means that the students can be set more interactive questions.

Using the example elsewhere

This type of activity can be used for producing interactive lessons where you want your students to explore the patterns in data, rather than the actual mechanism of producing the data. A similar activity might be to produce a spreadsheet that constructs a graph as students enter distance–time data. Another activity might be to produce a simulation of current–voltage graphs.

5.5 Self-checking for students

Chemical calculations

In the field of chemistry, calculations are an area that students persistently find difficult. We can use Excel to provide a self-test, i.e. an opportunity for students to practise calculations on their own. There is also a nice logical progression to chemical calculations, starting with determination of molecular masses, moving on to determining number of moles and then the concentration of solutions.

The exercise in this example could easily be carried out by students using pen and paper, but one of the great uses of Excel is to generate formulae that will allow students to check their own results. This allows students to progress at their own pace.

Students use a spreadsheet as shown in Figure 5.10 overleaf to perform calculations involving a number of different compounds (column A). They calculate the relative formula mass of each (column B), and correct answers are indicated in column C. They can go on to calculate numbers of moles and concentrations of solutions.

In order for this to work, the spreadsheet has been programmed with values of the relative formula masses and has the necessary formulae to calculate the other quantities. The exercise could be split into sections and each placed on a different sheet in the same workbook, enabling the students to practise one activity at a time.

Formula	RFM	Correct?	Mas we	Number	(Correct?	Volume of solution (dm3)	Concentration (moldm-3)	Correct
NaOH	40	Correct	10	0.25	Correct	0.5	0.50	Correct
KBr	119	Correct	23	0.19	Correct	1		Wrong
HCl	36	Correct	72	2.00	Correct	3		Wrong
CaF2	78	Correct	200	2.56	Correct	0.25		Wrong
CuSO4	159	Correct	15.9	0.10	Correct	0.04		Wrong
HNO3	63	Correct	126	2.00	Correct	40		Wrong
SO2	64	Correct	256	4.00	Correct	0.4		Wrong
CO2	44	Correct	110	2.50	Correct	5		Wrong
Ca(OH)2	74	Correct	3.7	0.05	Correct	0.1		Wrong
Al2O3	102	Correct	51	0.50	Correct	0.2		Wrong
NaHCO3	84	Correct	42	0.50	Correct	0.05		Wrong
Ca3PO4	215	Correct	430	2.00	Correct	0.625		Wrong
MgO	40	Correct	2000	50.00	Correct	0.2		Wrong
Ca(NO3)2	164	Correct	820	5.00	Correct	4		Wrong
Al2(SO4)3	342	Correct	125	0.37	Correct	9		Wrong

Figure 5.10
The working spreadsheet for chemical calculations, using Excel.

Why it's worthwhile
The activity described above allows students to work at their own pace and to discover for themselves whether they have obtained the right answer, without having to wait for teacher input before they move on. It also therefore frees up some teacher time to assist those who are struggling with the task or to provide extra extension work for others.

Using the example elsewhere
Any activity that involves a calculation can be checked using the technique described. However, it needn't be restricted to numerical answers and can be used in any situation where the answers to a question are specific and require no interpretation.

5.6 Linking to multimedia

Investigating a chemical reaction
Spreadsheets can be set up to model the changing conditions of a reaction. This example uses the *Multimedia Science School* (PLATO Learning) tool 'The Haber Process' as the source of simulated data (Figure 5.11). This tool is very easy to use. By clicking on the different temperatures and pressures the conditions can be changed easily. Alongside the tool, values are recorded in Excel and used to plot a graph showing how changing the conditions affects the yield.

Figure 5.11
'The Haber Process' from Multimedia Science School.

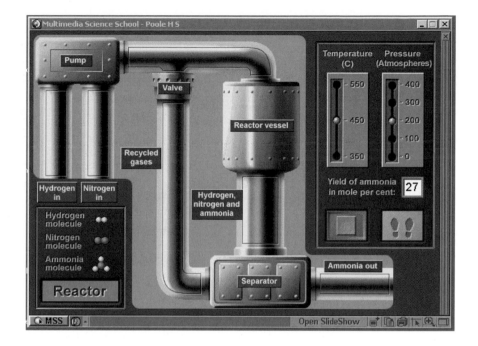

Students are provided with a spreadsheet in the form of a blank table, accompanied by a blank graph. They collect data from the multimedia tool and enter it in the table. Excel then produces the graph of their data (Figure 5.12). The activity so far is very straightforward; students simply manipulate the *Multimedia Science School* program and transfer numbers into Excel. In order to extend the activity and encourage students to think about the results they have produced, there is plenty of space on the right of the spreadsheet to add some questions for students to answer.

Why it's worthwhile

The Haber Process can be difficult for students to visualise and the teaching of it can be fraught with difficulties. Use of a model like the one described makes visualisation of the process much easier. Use of the spreadsheet enables students to experiment with 'real data' and develop a sense of what is happening as the conditions change. The fact that the spreadsheet is set up in advance so that the graph appears as the data is entered helps students to immediately distinguish between the effects of temperature and pressure.

Figure 5.12
*The resulting
spreadsheet with
data completed
graph drawn.*

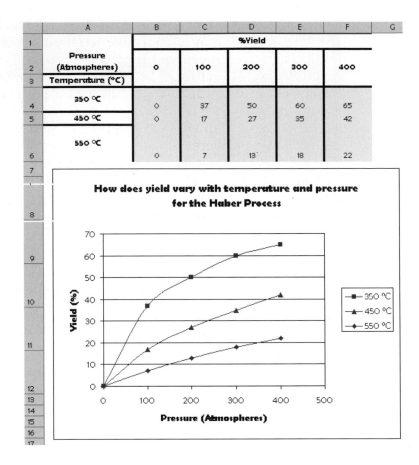

	A	B	C	D	E	F	G
1			**%Yield**				
2	**Pressure (Atmospheres)**	0	100	200	300	400	
3	**Temperature (°C)**						
4	350 °C	0	37	50	60	65	
5	450 °C	0	17	27	35	42	
6	550 °C	0	7	13	18	22	

How does yield vary with temperature and pressure
for the Haber Process

Using the example elsewhere

This use of spreadsheets to model a 'real world' situation can
be used wherever multiple sets of data require graphing for
interpretation. For example, I have used it successfully to
model the current–voltage characteristics of filament lamps,
diodes and resistors at constant temperature.

◆ *Conclusion*

It is important to be clear why you are using ICT to handle
data in science lessons. It can provide a new approach to old
problems and this can be advantageous in the classroom. It
also allows those students who have difficulty in manipulating
the numbers or in drawing graphs full access to the science
behind the data. The power of applications like Excel means
that it is possible to use huge data sets, for example in
population studies.

The most significant drawback of these approaches to data handling is the cost in teacher time required to set up and develop these resources, and the first few times they are used in the classroom. It is at times like these that a supportive department is essential, so that members of staff can collaborate and share resources.

From the point of view of use in the classroom, there will be the inevitable problems of hardware deficiencies and random software crashes. Cost is less of an issue with data handling than with other ICT applications because, with the exception of the final example above, most uses require only generic software that is readily available and which should function with few problems.

◆ *Resources*

Census At School
   www.censusatschool.ntu.ac.uk
Multimedia Science School
   PLATO Learning www.platolearning.co.uk

Modelling – a core activity

Ian Lawrence

♦ Introduction

Good science makes good pictures of the world. At their best, these are dynamic pictures, accounting for changes over time, showing what changes and what does not change. A modelling tool is a piece of software that allows you and your students to develop such pictures.

Comparison of these pictures with the real world decides their value. A good match means you have done a good piece of science. This can be hugely satisfying. Activities with modelling software are designed to allow students this pleasure. They let the students write the rules that define what the system does. There may be valuable mistakes made. As with any open-ended activity, the skilful teacher guides progress through the activity, so that lessons can be learnt from false starts, and successes of many kinds can be appropriately celebrated.

The same tools can be used by the teacher to guide more convergent activities, perhaps constructing a pre-practised model with the class, or to source or prepare a particular model before the lesson starts. Even at this end of the spectrum of use there are choices to be made. With the completed model you can explore just the inputs and outputs, so using the modelling tool to produce simulations, or you can explore the structure of the thinking represented in the model with the class, so having a powerful and public way of sharing your thinking. And that is what modelling software provides: a set of tools for expressing your own thinking, so making it clear, especially to yourself.

Now you no longer have to be able to program in order to supplement word processing and drawing diagrams with a more dynamic way of saying things with the computer. Modelling tools provide carefully controlled access to a wide range of expressive powers, so that you, and the students you teach, can create and share your own narratives. I believe most fun is had when students really get the chance to express themselves, so now let's see how to achieve that.

6.1 Thinking about tools and activities

Making sense of things with science is often a question of finding a pattern. In the sciences, I think we can usefully reduce everything down to patterns constructed around two kinds of building blocks: *systems* and *objects*. So there are two kinds of structures for which we need to provide modelling tools. I think it is both possible and helpful to characterise modelling as building with one of two different kinds of construction kits. Figure 6.1 shows these two 'kits', one of which represents thinking about systems, the other thinking with objects. *VnR* and *WorldMaker* are two free examples of such tools (see page 128). Both are suitable for substantial classroom use.

There are, of course, a number of ways of showing how things are related, for example:

- words
- diagrams, especially graphs
- mathematics.

(Mathematics is especially good for representing patterns, so a vital question is what mathematics can be drawn into the teaching, showing a sensitivity both to what has already been done, and to what could be done. This boundary between mathematics and science has proved very fruitful.)

Figure 6.1 *Modelling in science clarifies thinking about **systems** and thinking about **objects**.*

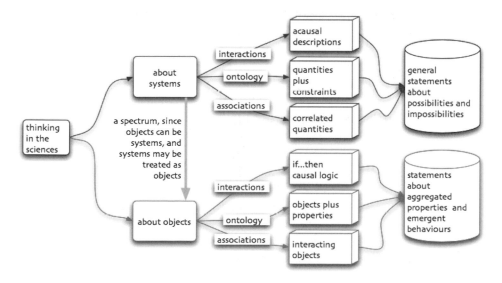

Modelling adds to the list, drawing on all of these ways of expressing thoughts, but adding another in allowing students to see their thoughts played out in an imaginary, yet public, world.

Modelling has an increasingly prominent role across all research sciences; this prevalence suggests that at school level a fair representation of science should try to expose students to this approach.

Choosing activities

Modelling activities with an appropriate computer tool can encompass a range of different ability levels, ages, topics and time slots within lessons. In this chapter a few good modelling activities have been chosen, and a varied selection from their many uses given, just to allow you to think about the possibilities.

The main purpose of activities centred on modelling with computers is that students are able to make dynamic representations of the science, which change over time or in response to user input, but, crucially, where students can write the rules. So the tools that support these activities are tools that allow students to express themselves. As a consequence of this they are also tools that are very good for sharing thinking between students, as well as making the teacher's thinking explicit to the students.

6.2 Thinking with variables

The following examples, all built using *VnR*, show what is possible when you give students access to tools that allow them to construct networks of variables, with the links made by relationships between those variables. That might well not be the best way to present the software to the students, but you can see that it does make explicit the very foundations on which science is built. For those of you with long memories, we have recreated the analogue computer – only this time in software!

Photosynthesis – brainstorming a summary

Imagine you are summarising photosynthesis, perhaps with a clever class. A brainstorm correctly identifies the factors affecting how much photosynthesis occurs and what the outputs are. Instead of simply listing these on a board, you see the chance to capture the relationships between these variables and then to explore the interconnections between them – see Figure 6.2. Alongside the build-up of the model, you reinforce the relationships, emphasising the meaning of each connection. You'll also be able to explore the important point of a limiting factor, in deciding to combine the relationships with a multiplication sign – so that 'no water' leads to 'no photosynthesis', however sunny (and vice versa).

Figure 6.2

Inputs and outputs from photosynthesis are clearly shown in this VnR model. Calculations are live, so that changes in any input are reflected in outputs, allowing the live exploration of "What if?" questions.

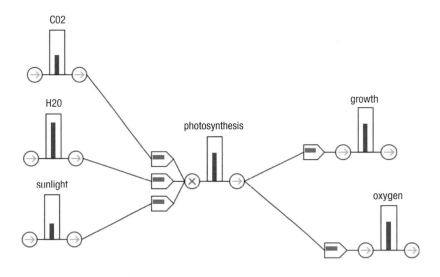

You can also take the model further, showing how the increased area of the leaves leads to more photosynthesis, so leading to an introduction to feedback (Figure 6.3).

This activity might require 5–15 minutes of classroom time. Other areas that can be approached in this way include energy resources and scientific investigations.

Figure 6.3
*Inputs and outputs
with feedback.
The growth sets
the rate at which
the plant
increases in size –
it just keeps on
growing. This
increased size
leads to more
photosynthesising.*

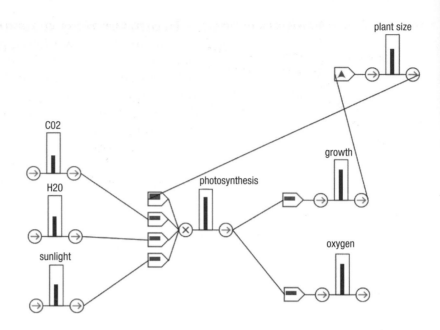

A simple ecosystem – a model of organism numbers

This is an example in which students build a model to explore a
partially understood system. You have studied food webs and
want to look at how the numbers in a food chain come to
stabilise. This is quite a difficult argument to get right, almost
impossible without the right tools. The scenario is that you
have been through the food web once, perhaps by a
diagramming exercise, or by cutting and pasting statements into
order. You are looking for a new way in, to reinforce and extend
what the students already know. You challenge the suggestion
that the stabilisation of the numbers of different organisms in
the system can simply be down to magic or some unspecified
process. So the scene is set to build up a model. Because of
their prior experience with this topic you judge that the
students should be able to build up their own model. For this
to be successful they will need not only to be quite familiar with
the topic but also to be competent with the modelling tool.
You will need to provide appropriate support sheets, marking
out mileposts on the way to the completed model, but without
leading students so much that they are making no decisions at
all. So the management of this process will be crucial.

Figure 6.4 shows some possible stepping stones on the way
to a completed model of rabbit numbers, together with things
students might say at each stage.

Figure 6.4 *Possible models of factors affecting rabbit numbers in an ecosystem, and students' statements.*

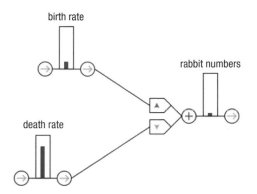

"Rabbits breed, so rabbit numbers keep on growing."
"So let's put a 'grow like me' link in there."
"And they die."
"So we'll put a 'shrink like me' link in there, OK?"
"Right, let's try it, see if we can keep the number of rabbits the same."

"More food makes the rabbits breed more... put a 'follow me' link in then."

"Some rabbits get killed. Things eat them."

"Predators- more predators, so more deaths. Another link, like the births."

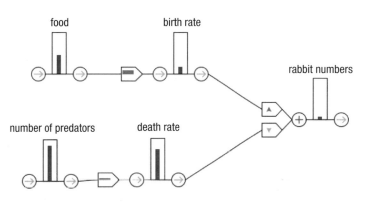

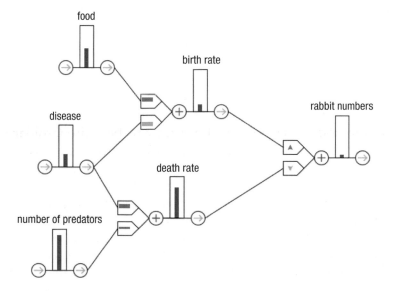

"What about disease?"

"Yes that would stop them breeding so much. More disease... less breeding."

"So that's a 'follow me' link then, but flipped."

"Disease will also kill them – so a higher death rate."

"More disease means more death."

continued

Figure 6.4 *continued*

"But the numbers of rabbits will also affect the food."

"Same number of rabbits means the food goes on getting less and less."

"So another 'grow like me' link, then, but flipped."

"And the number of predators, they'll go up too, if the rabbits go up."

"They'll keep on going up and up, if the rabbits stay the same."

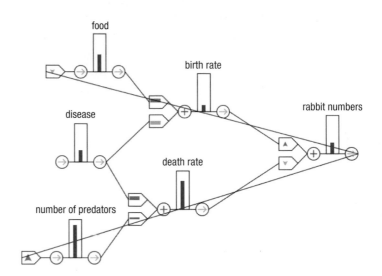

Here the focus is on the dynamic evolution of the model, seeing how patterns in the numbers evolve, and how these changes come about in detail. The shape and structure of the system as a whole is important, but so is the chance to take the model apart. This deconstruction is vital for seeing how the balance in numbers comes about.

This activity might require 30–60 minutes of classroom time, depending largely on the amount and focus of the support. Further examples that can be modelled in this way include the falling parachute, vehicle stopping distances (see also section 5.4) and rates of reaction.

Cooling things down – building up a model together

Here the idea is to jointly explore some thinking about energy transfer from warm things to cooler things. Some careful talking through of each step, so generating a shared understanding of what is to be represented, then careful ordering of each of these stepping stones, can give quite a detailed understanding of the thermal interactions of an object with its surroundings. As an introduction to the students using the modelling tool, rather than just watching you use it on the interactive whiteboard, you might decide to work through the model with them. To ground it in the 'everyday', have a cup, a teabag and some milk to hand.

Start by making up a cup of tea to a comfortable temperature. Making it in front of the students, adding the hot water and the milk, then asking how the quantities of these two contribute to the overall temperature, leads naturally to an expression (Figure 6.5) – only a short discussion is needed.

Figure 6.5

More hot water leads to the tea becoming hotter; more milk leads to the tea becoming cooler. So water adds to the temperature of the tea in a positive sense, and milk in a negative sense.

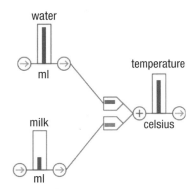

Then you need to compare the temperature of the tea with the surrounding temperature, building on the students' knowledge that things warmer than their environment will cool down. This uses the same structure in the model as before, so showing how the same piece of thinking can be re-used (Figure 6.6).

Figure 6.6

The tea temperature is compared with the surrounding air temperature. The tea temperature adds to this comparison variable, and the air temperature contributes in a negative sense.

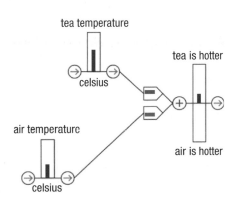

Then the two chunks of model can be combined, so that the tea temperature is common to both (Figure 6.7).

Figure 6.7
The results from the first comparison are fed into the second comparison.

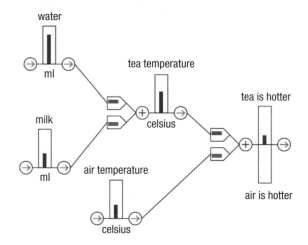

Now the knowledge of the class that 'something with a temperature higher than its surroundings cools down' can be built into the model. Drawing on this leads to 'when the tea is hotter the tea keeps on cooling down'. So a feedback loop is added (Figure 6.8).

Figure 6.8
The feedback loop literally feeds back. When it arrives it contributes to the rate at which the tea temperature must drop, signified by the downward arrow.

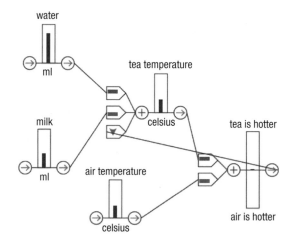

Then the model is essentially complete. Several runs with a commentary can show how the tea cools down from different initial temperatures, and how changing the ambient conditions affect that cooling (Figure 6.9).

Figure 6.9   **a)**

The tea temperature gradually approaches the air temperature. The graph is produced by the model and auto-scaled. A higher tea temperature (partly obscured by the auto-scaling) and a lower outside temperature produce a greater rate of cooling.

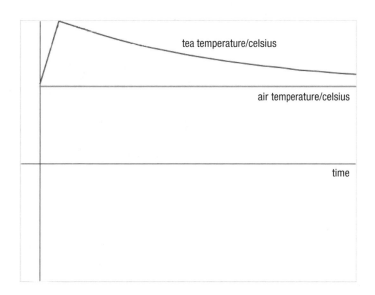

b)

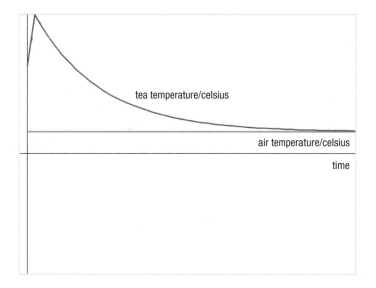

Allow 20–30 minutes of classroom time for this cooling model. Further examples that can be modelled in this way include ecosystems, parachutes and terminal velocity, stopping distance, and respiration in plants.

Geology – using the modelling tool to sharpen up thinking

In the examples so far, the modelling tool has been the focus of the activity. Here is an example where the tool supports another activity, by giving the students a chance to clarify their ideas. The context is an investigation in earth science, where the students are looking at lava bombs.

Here the challenge is to find sensible connections between the speed with which the ejected lava bomb hits the ground and the area it covers on impact near the volcano. This involves chaining together several steps of the argument, so as to connect the input variable (speed) with the output and to show the assumptions explicitly in all of the intermediate stages. The student needs to be able to alter the relationships connecting the variables as well as their values.

In this example (Figure 6.10) a student develops the model through several iterations with feedback from the teacher.

Figure 6.10 *Reading the model back: "A lava bomb hits the ground with a particular velocity. I can work out the height from which I would have had to drop the bomb to get this speed. This speed also determines the kinetic energy available to reshape the bomb. That in turn fixes the area."*

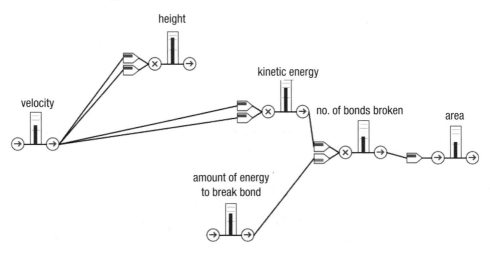

With the aid of this model the student is able to refine the focus of the investigation and to make a much more precise statement of what they set out to investigate, and a much more precise statement of what they expect to find.

So here the student is reasoning about processes in order to build a model that is more coherent and more extensive than could be built unsupported by the computer. In particular, the modelling tool encourages the student to make each step of an argument explicit. The exact meaning of the variable and relationship building blocks supports precise thinking, and it is hard to stray. There are several places in the science curriculum where this kind of multi-stage argument needs you to hold several steps in the head; linking them in the way that this tool allows permits you to explore the connections across stages much more thoroughly. In investigating unfamiliar situations where the answer is not known with certainty beforehand, this kind of semi-quantitative, multi-stage reasoning is very helpful as a guide to thoughtful action – and not only in the sciences.

Allow about 20 minutes of classroom time for this activity, plus plenty of time for reflection. Further similar activities include exploration of stopping distance, rates of reaction, respiration and cooling.

6.3 Thinking with objects

A different kind of tool is one that allows you to reason with large collections of objects that interact by following rules. This represents a second fundamental way in which scientists reason about the world. Again, you can provide access to the basic building blocks, with which students can express themselves.

Ecosystems – seeing the objects as living things
One place to start with these kinds of tools is by building simple ecosystems. You could start with a food chain or web of only a few stages, building up the model step by step. There are several key stepping stones to build in. Starting from the science, you would probably want rules to capture at least being eaten, reproduction and movement, for at least some organisms. Then you can see how complex patterns for the whole system emerge from simple behaviour for each individual in the system, as captured by these same rules.

Again here there are decisions to be made about the degree of support to be provided. You might choose to start off with a blank screen, and a file containing the individual objects, leaving the students to build the rules defining the behaviour of these objects. Frequent checking of the rules governing behaviour that the students code is advisable. You might provide a few rules completed, or even all of the rules completed, but unnamed, so that the students have to recognise what behaviour these rules are supposed to represent in the system being modelled.

The completed model for a simple ecosystem might look like that shown in Figure 6.11, as realised in *WorldMaker*.

The crucial rule to get in place early on is the one for movement, as this brings the model to life, showing students that they are making progress. Consuming other members of the system and reproducing can follow, in any order. Adding features such as death from natural causes provides even more complexity to the model without changing the essence of what is being captured, so can follow even later.

Figure 6.11 *An overview of a simple ecosystem in WorldMaker, showing the rules for the rabbit.*

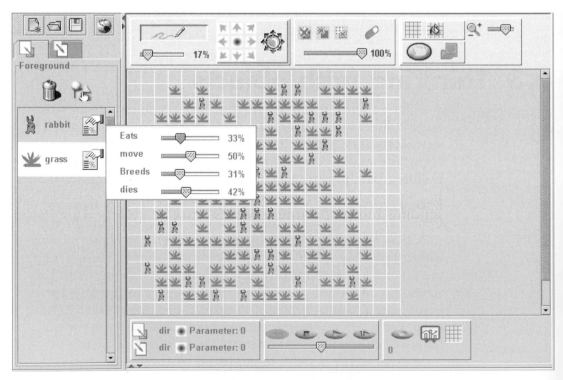

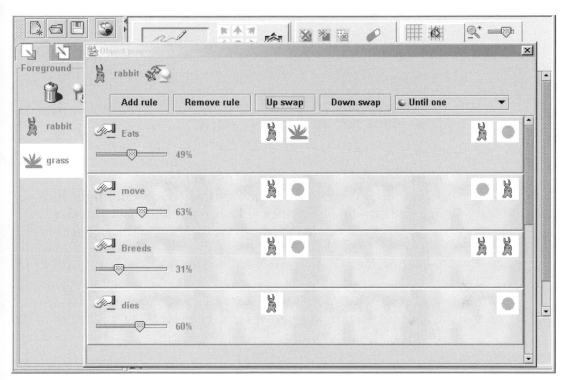

Figure 6.12
Details of the rules for the rabbit. Both the chances that each rule fires (expressed as percentages) and the order of the rules are important.

Focusing on the basic ecosystem with its simple behaviour rules (Figures 6.12 and 6.13), and discussing how to judge the success of the model, gives plenty of opportunity to draw out points about the nature of scientific enterprise – in that it tries to represent a few aspects of nature well.

Again there are important decisions to be made as to how these steps in the modelling process are staged and, depending both on the aptitudes of the class and on their facility/familiarity with the modelling tool, how complex the model is at a point where you judge a suitable end point to be. The wider environment, beyond the screen, is vital in guiding and supporting students' efforts, while leaving them enough space for the efforts to be recognisably theirs.

Allow about 30 minutes of classroom time for this activity. Further examples that can be modelled in this way are rates of chemical reaction and the spread of a disease.

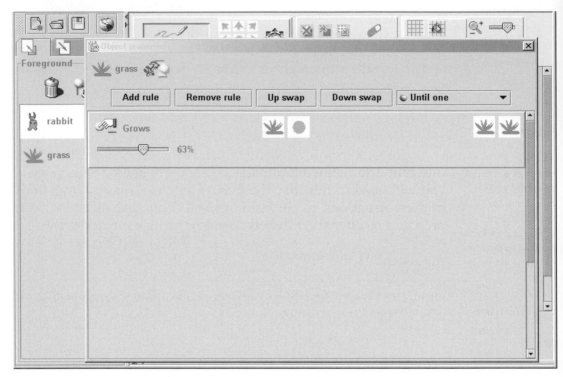

Figure 6.13
The grass has a fairly passive role, so needs fewer rules.

Particles – the simple behaviours of gas molecules

Another simple situation which has some consequences that may surprise and inform, together with programmed rules that illuminate the science, is the movement of gas molecules. The behaviour of particles is an important topic in early studies of chemistry, and having modelled this behaviour gives the class a shared experience to draw on as they come to a clearer understanding of particles.

The essence of this model is that no special directionality is required to drive a change – some properties of this state of matter emerge just from the random movements of the molecules. There is nothing purposive about this movement – purely local behaviour is sufficient to account for diffusion and for other macroscopic material properties. This seems counter-intuitive to students, and so building a system where you control every aspect of the rules allows you to explore this strange, but causal, behaviour. Working from the behaviour of a single particle can make the bulk properties seem much more intelligible. Completely random movement of particles on a grid is easy to simulate in a number of systems. Figure 6.14 shows the rules in *WorldMaker*.

It may take students some time to get this seemingly simple system right, but I think it is time well spent, as their understanding of the rules is of the essence here. The challenge might be that the rules seem too simple!

You might like to try other rules for movement as well, to compare and contrast, asking which is the best for a number of different arrangements of particles. You are likely to find that you will need to rehearse the behaviour that you *expect* to find in these situations, as the students will otherwise often be prepared to alter their beliefs about what happens to fit the model. In fact continuous and close comparison between the model and the situation being modelled is vital here.

Figure 6.14
The three parts of the rules show the detail in each rule: what the object must look for, what class of action it can take, and the detail of that action.

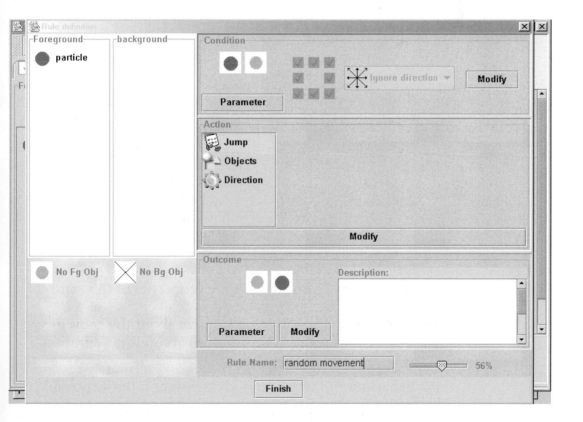

Figure 6.15

A simple rule gives some chance that two different particles will react on meeting. Following the change in the number of particles reacting can model changes in reaction rates.

A universal behaviour pattern for particles is best here – you can now bring this out explicitly. This allows a grounded discussion of ideas and evidence, drawing on the experience of playing with the models. However, this may not be suitable for all. Some will express themselves more fluently with the modelling tool than in oral communication; it would be an underuse of the medium to try to reduce too much of the experience to words.

The same ideas can be extended to deal with chemical reactions, showing how the movement of the particles, and hence the temperature, affects the rate of reaction (Figure 6.15).

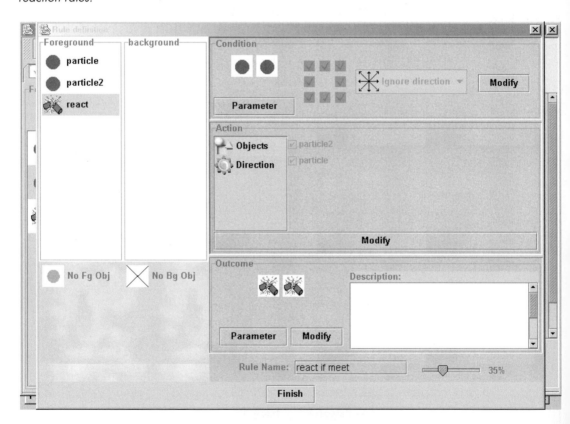

Thinking about how to set up the tool to accurately represent the behaviour of the particles gives insight into the possible behaviour of the particles in a wide range of contexts, and it gives an experience to refer back to as students continue with the study of particles. Even if the models are not developed to cover more complex situations later, this modelling experience can be built into those later lessons in such a way that the fundamentals remain in focus.

Allow 30–40 minutes of classroom time for this activity. Further examples using the same ideas include the spread of organisms across an area and the irreversible mixing of liquids.

Radioactivity – the same wherever you are, whatever you are doing

Figure 6.16
The independence of action of the nucleus is the key point here.

Here you will want to emphasise that nothing that goes on around an unstable nucleus affects the rate at which it decays. So the essential rule for all unstable nuclei is that they turn into something else with a fixed probability that does not depend on their location, neighbours, temperature, or any activity on their part. Figure 6.16 shows that rule in *WorldMaker*.

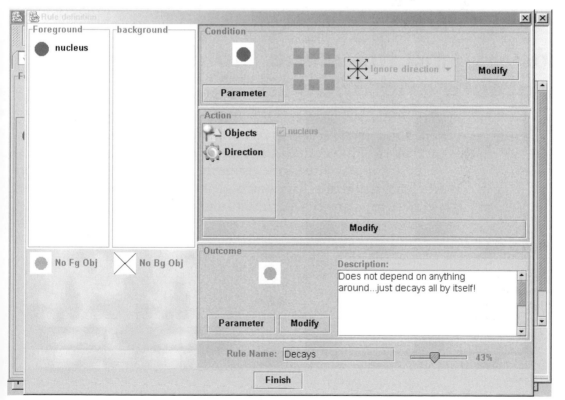

The particle's location and neighbours are of course related to possible interactions with surrounding particles, so accounting for chemistry and the whole of macroscopic physics, except for temperature, which you can correlate with the movement of the particle. This persistent internal checking, whatever the other rules, is the key. You might say, "It's the number one rule." Certainly you'll need to put it first in the list to be certain that it gets checked in every tick of the program's clock, unless you change the order in which the rules are checked. Taking this precaution will make sure that the half life is truly constant, whatever other rules you add, so long as you do not write another rule that destroys the particles.

It is easily possible to extend this to daughter products, and to follow the changes in the numbers of all the members of a (short) decay chain.

Allow 30–40 minutes of classroom time for this modelling activity. Any other exponential decay can be modelled just like this, for example the absorption of light by successive plates of glass.

6.4 Using the same tool for data capture and analysis

The tools used so far have avoided the use of algebra, as I do not think many students can reliably reason with it before the age of 16. However, *Datalogging Insight* (see page 128), which you may already be using for datalogging, does support the use of algebra. It is worth giving an example of its use here so that you can see whether it might be useful to you.

Figure 6.17
A simple model to continuously add increments of distance (steady speed running for a short time interval) to find the total distance covered.

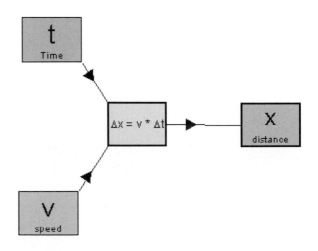

Figure 6.18
A graph of distance–time for a steady speed, produced by Datalogging Insight *with the data generated by the model in Figure 6.17.*

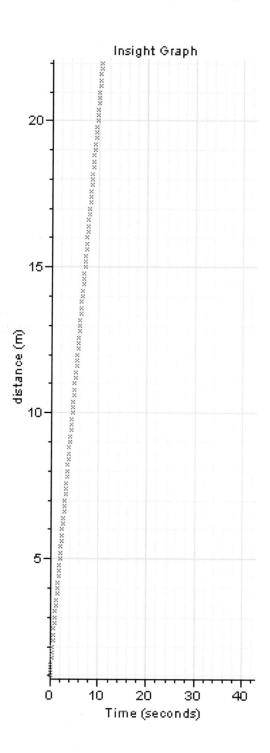

Modelling motion

One use for the embedded modelling tool is to generate a stream of data, just like that provided by a datalogger. The same analytic tools can be used to compare the two streams of numbers, looking for similar patterns. The illustrations here show how the modelling tool can be used to develop a series of models to represent motion.

You might start with something moving at a steady speed. To work out how far it has gone, you need only make a few connections: a constant and the steady ticking of the clock combine to increment the distance covered, with this calculation defined in the linking box (see Figure 6.17, page 120). The graph in Figure 6.18, page 121, is obtained.

Then you can change the constant to a variable, add a new constant, and define one more calculation to show how steadily accelerated motion can be represented (see Figure 6.19). This produces a new graph, showing steadily accelerated motion (Figure 6.20).

Figure 6.19

A two-step model. The first step continuously adds increments of speed (steady acceleration running for a short time interval) to find the current speed. This speed is then fed into a second stage, with the model now continuously adding increments of distance (speed running for a short time interval) to find the total distance covered.

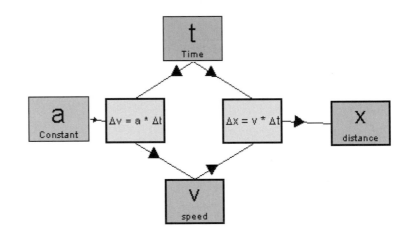

Figure 6.20
A graph of distance–time (upper line) and speed–time (lower line) for steadily accelerated motion, using the data generated by the model in Figure 6.19.

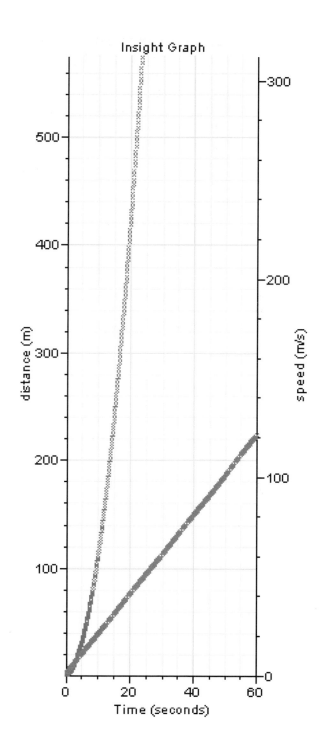

As an end point you can get most of a 16-year old's understanding of Newtonian mechanics into a single model, but care is needed with this compression. See Figure 6.21, which incorporates Newton's second law into the model. This produces a similar graph, again showing accelerated motion (Figure 6.22).

Figure 6.21
The two-step model, further extended to incorporate Newton's second law.

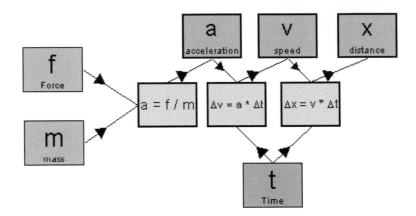

You will need to decide whether building these kinds of models helps your students to develop a better understanding of the relationships that are represented.

Figure 6.22
A graph of distance–time (upper line), speed–time (middle line) and acceleration–time (lower line) for accelerated motion, using the data generated by the model in Figure 6.21.

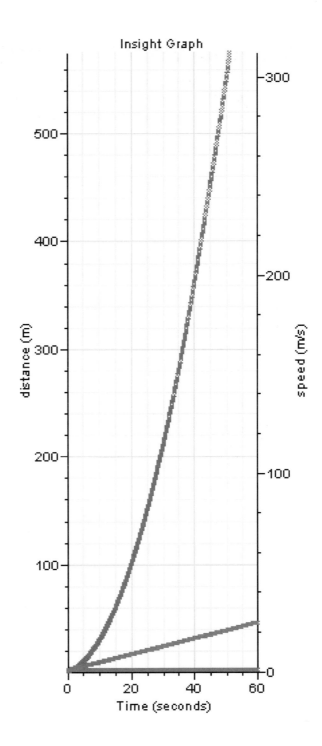

Insight Graph

6.5 Moving on with algebra

If you have some students who can reliably work with algebra then the best tool I know of is *Modellus*™ (see page 129). Even if your classes in their entirety do not enjoy it, it will be appropriate for some, including perhaps the teacher! It can be used at this level to produce simulations for producing demonstrations and other interactions. The illustrations here (Figures 6.23 and 6.24) show two examples from many that could have been chosen.

Figure 6.23 *A simple model with teacher-friendly animation, showing how force and acceleration are linked for a fixed mass (changed part way through in this trace).*

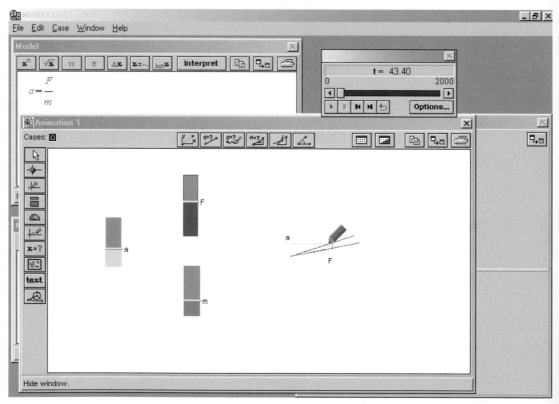

Figure 6.24 *Another one-line model, showing the connection between mouse movement (in one direction) and the velocity vector: as the object is dragged, so the vector is shown.*

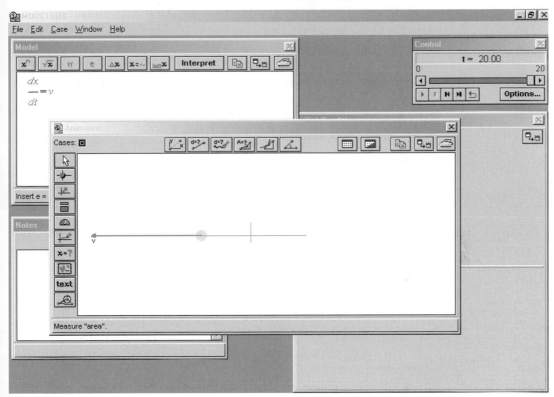

6.6 Choosing tools: accessibility and availability

There are only a few modelling tools (perhaps we could dub them 'world processors'?) appropriate to the abilities of students in the 11–16 age range. So it is not the case that one can write activities, as one would perhaps for any word processor, spreadsheet or drawing package. The activities suggested in this chapter presuppose these key things in a modelling package:

- low algebraic demand
- clean representations of the key building blocks to think with
- multiple ways to represent the outputs of the model
- maximum simplicity and minimum choice appropriate to the task
- the chance to interact with the model while it is running.

VnR and *WorldMaker* are two good examples of such tools, and have been used to generate many of the examples in this chapter.

VnR was developed as a part of the ASE's *Science Year* (2002). It is therefore free, and is available on the CD-ROM accompanying this book.

WorldMaker has been developed from a much earlier version; the current version is available free online (see page 129).

Both have been used here to present a small, but representative, selection of activities from among the many areas in which you could use these tools.

Other packages to consider

Datalogging Insight (the fourth version of the *Insight* software), used in section 6.4, is a commercial package available from Longman Logotron (www.logo.com).

Both *Stagecast*™ (www.stagecast.com) and *AgentSheets* (www.AgentSheets.com) are probably worth later investigation but are expensive, so might best be left for a while until you have some experience with the free packages against which to judge their merits and demerits for your purposes. Neither is designed for modelling as such, and both fall into the style of thinking best characterised as 'objects and properties'.

For further exploration within the 'variables and relationships' paradigm, both *Modellus*™ and *Stella*® are good but probably inappropriate for use by students at KS3 and 4 (ages 11–16), unless students are algebraically competent, or unless you are prepared to take time to make them comfortable with algebra.

Modellus™ is a good tool for teachers to work with (see section 6.5), but requires that the model be expressed in algebra. It is excellent for post-16 studies. It is available for free download (see page 129) and the current version is provided on the CD-ROM accompanying this book.

Stella® (www.hps-inc.com) is very expensive, and the surface friendliness is replaced by the need to write Basic-like statements to define key relationships.

A brief word on spreadsheets: these are effectively free, since now few computers come without them, but they impose an extra layer of translation between what you want to say and how you can say it. This extra layer does not make them good for expressing the science, so I strongly recommend against asking students to develop their own models using spreadsheets at this level.

◆ *Conclusion*

Why use modelling tools?

The essence of the activity of modelling is that you represent the structure of the thinking in an area. It is a means of expressing yourself, showing how you think things are connected together. Often the teacher does the modelling while the students just see the simulations. In this chapter, I hope I have encouraged you to let the students have a go at the modelling. This will allow them to express their own thinking, and to share it.

The challenge for you, the teacher, in managing modelling in the classroom lies both in constraining and in challenging. You need to establish appropriate stepping stones, to allow students to know when they are on the right lines, both to celebrate partial success and to recognise successful completion.

Resources

Datalogging Insight
www.logo.com
Modellus™
http://phoenix.sce.fct.unl.pt/modellus
Also available on the CD-ROM accompanying this book.
VnR
http://mise.bham.ac.uk
Also available on the CD-ROM accompanying this book.
WorldMaker
http://worldmaker.cite.hku.hk/worldmaker/pages/home.htm

7 The effective use of the Internet

Dave Pickersgill

♦ Introduction

This chapter explores practical and effective ways of utilising the Internet when teaching secondary science. Aspects covered include search techniques and downloading, but the prime aim is to alert you to some of the multitude of useful resources available on the Internet and to suggest ideas and examples that will enhance both your lesson preparation and your teaching.

7.1 A view of the Internet

The Internet can be likened to a giant, ever-growing library in which the contents are totally disorganised! Some catalogues exist but none are complete. Yet this library will contain the information you want, at the level you want. The only difficulty is in locating that information, in assimilating the diversity of sources and then in assessing their reliability. Your students will require guidance on the quality of the information available. On the Internet, there are few guarantees about what is accurate and impartial. Simple methods of evaluating websites include (i) checking the date of last update, (ii) identifying who produced the information, and (iii) obtaining factual confirmation from elsewhere.

It is essential for any Internet user to develop a clear search strategy: a strategy that will evolve with practice and will become more and more sophisticated. An activity that involves searching can be part of a natural circus of events – a research-based aspect of your normal teaching.

It is possible to use the Internet productively with students but it is important to provide clarity of task.

- Exactly what is the student looking for?
- Will the student achieve 'success' relatively easily?
- Will the task be sufficiently stimulating?

After a number of such positive experiences, a user will gain in confidence and be willing to persevere when their initial search produces '15 000 hits'. Perhaps there is a need to narrow my search terms? – 'wind power, UK' rather than, simply, 'wind'? The development of such search techniques encourages lateral thinking and stimulates reflection on teaching and learning, in addition to encouraging collaboration and support among your students.

You could take a very structured approach by previewing websites and devising activities around them. Your students will then obtain exactly what they need, and can work on their task towards an expected outcome. You could develop a simple template (see Figure 7.1), around which your questions could be posed. However, teachers do not have the time to undertake such detailed preparation for all the opportunities when it is possible to access the Internet. Students need to develop their own search skills, but can be directed towards appropriate sites as they commence their searching.

When using the Internet in the classroom, you may find things not going to plan. The Internet connection or server may be so busy that pages take an age to appear on screen, or the document you bookmarked has been renamed, deleted or moved. Opening a new browser window lets you explore elsewhere, but this is unlikely to help you complete your original task by the end of the session. The Web is never the same two days in a row; this very strength and flexibility can sometimes prove to be a weakness.

Figure 7.1
A teaching template for work using the Internet, produced using a word processing package.

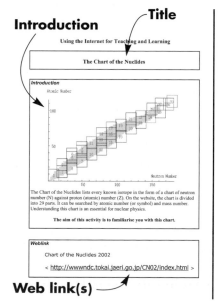

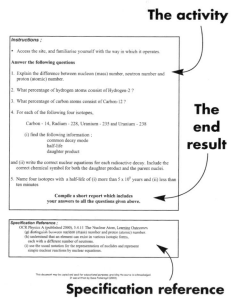

Box 1. Accessing a website

A URL (which stands for Uniform *or* Universal *or* Unique Resource Locator) is a location on the Internet, generally a web page. Thus Internet users 'visit' web pages by 'entering' URL addresses into their browsers. The URL for a web page is generally of the form:

http://www.something.somethingelse/directory_details

Here are two, for example:

Google™,   http://www.google.com

Sheffield College Weblinks,   http://www.sheffcol.ac.uk/links

In order to access a web page, you should open your web browser (for example, Internet Explorer, Netscape® Navigator or Opera) and type the URL into the address box (the http:// before the www. is not needed), then press the return key. If you are online, the browser will then 'visit' the website you have selected. Using the right mouse button on your PC, it is also possible to copy and paste a URL into the address box of your web browser.

7.2 Using the Internet with your students

The web addresses of all the examples given are listed at the end of the chapter (pages 142–3).

Astronomy and space – images and interactivity

There are many useful websites concerned with astronomy which are suitable for use with your students. These vary from many excellent image-laden sites to sites that allow you to use large astronomical telescopes in real-time. The former has the advantage over paper-based material of allowing you to manipulate images in order to produce material easily, while the latter offers opportunities which in pre-Internet days would have proved prohibitively expensive. The following three examples include one excellent image repository and two sites that can provide ideas for suitable activities.

Anglo-Australian Observatory

This site acts as a gateway to a large and unique collection of wide-field astronomical photographs. These are all available for download in a range of standard sizes up to 100 MB. This is an excellent resource which can provide the image you want.

See the International Space Station

The International Space Station has now been orbiting the Earth over 15 times a day since the year 2000, yet how many of us have actually seen it? Spotting it is not as difficult as it might seem, providing you know in which direction to look, and when. This site gives all the information needed to find the biggest and most powerful spacecraft ever built as it passes over our heads. This is an excellent activity for those long summer evenings or when you have taken students away for an overnight stay.

Faulkes Telescope

The Faulkes Telescope Project is building two large astronomical telescopes (2.0 m reflectors) for the exclusive use of schools. Users will then be able to book real-time half-hour slots on one of the two telescopes, situated in Maui, Hawaii and Siding Spring Observatory, Australia. The project will also provide supporting materials. By making use of the Internet, the Faulkes Project can offer a unique experience for our students.

The Environment – analyse the data

Environmental science offers many opportunities for your students to undertake detailed data analysis. The following sites provide up-to-date and detailed information, which can easily be used in the classroom. If you wish to progress further into data analysis by the use of detailed statistical techniques, *A New View of Statistics* (see page 141) offers the information you will require.

The Environment in your Pocket 2003

This is on the DEFRA site (the Department for Environment, Food & Rural Affairs). It contains data, notes, graphs and charts on a large variety of environmental issues, ranging from air quality, waste recycling and radioactivity to wildlife and public attitudes. It is an excellent source of up-to-date data. The site can be used in a variety of ways, for example for downloading directly to a spreadsheet for alternative display techniques (highlight, *copy*, then *paste*), or using the data to confirm ideas regarding the environment.

UK Pollution Data
This site allows you to select places around the UK and obtain hour-by-hour graphical pollution data for the place selected. The data gathered could be very useful in the classroom.

Neighbourhood Statistics
This site offers a multitude of information. You can view, compare or download statistics on a wide range of subjects including population, crime, health and housing. This can be achieved either by consideration of a local area or by subject. This information can be used by your students to compare their locality with the national picture or with the locality of their twin town.

Earth Science – experiments and information

SEED Science Centre
This is a public site where Schlumberger scientists and engineers share their knowledge and experience with learners and teachers from around the world. It includes links to a large number of suggested practical activities, suitable for upper primary and lower secondary school classes. Practical experiments suggested include a number of Earth Science activities: 'How round is the Earth?', 'Oilwell Blowout Simulator', 'Porosity' and 'The Absorbency of Rock'. Each activity includes detailed illustrated teaching notes, sample results and related links. The experimental notes are available in .pdf format. These could save you hours of preparation.

British Geological Society
This site includes downloadable screensavers which could be used to enliven your teaching in this area.

The Essential Guide to Rocks
This BBC site incorporates excellent animation and an interactive timeline featuring 4600 million years of continental drift, fossil life and rocks.

Revision

Brightsparks
Developed by CAST (the Centre for the Advancement of Science and Technology education), Brightsparks offers a very

large set of science revision quizzes suitable for both primary and secondary students. Hundreds of multiple-choice questions have been put together, each involving a picture and four answers to choose from. Teachers can keep track of how their students are progressing. Installlation involves a number of downloads, including Quicktime 4, although full instructions are given on this link. This is a 'must-have' resource.

Box 2. Downloading material

Given a project such as preparing a report or display, students can assemble the web-based material they need. They can view a page, highlight text or pictures and copy them into a word processor. Or they can save text, pictures or a whole web page as a file on disk. In this way they can treat web pages as raw data to mould for particular purposes. Most downloads may be accomplished relatively easily; for example, when using Internet Explorer, click the right mouse button. This will allow you to *save* the image or text file. Alternatively, you can highlight the text that interests you, then click the right mouse button to *copy* and then *paste* into your word processing file. Due to memory constraints, it is advisable to use the former technique when dealing with images. They can be saved and later inserted into your Word file (using the main menu, *Insert/Picture/From File*).

7.3 Teacher resources

Some little gems
The Internet offers many simple resources which are suitable for use by the teacher. They can often be downloaded and used in a variety of ways. A few examples are given here. Again, web addresses are given on pages 142–3.

Electricity Map
This consists of a single page, including a chart that plots current against voltage for a large number of electrical phenomena. The chart can be copied onto an OHP slide or handout. It can then be easily utilised as a lesson starter or as part of a plenary, while the teacher fires questions at the class.

Heart

This resource on the human heart, available on the accompanying CD-ROM, was produced by New Media (now PLATO Learning) as part of *Science Year*. It is possible to match the activity of the heart to the ECG displayed, compare the effect of different activities, demonstrate which sides of the heart pump the blood to the body and to the lungs, show how the blood oxygen level differs on the right and left sides of the heart, and much more. This is an excellent, free teaching aid which has a multitude of uses at various levels.

Figure 7.2
A view of PLATO Learning's Heart.

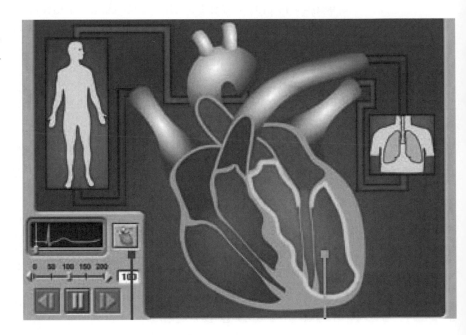

Powers of 10

This site is just what you need when discussing standard index form and powers of ten with your students. It is very impressive when projected onto a large screen. You start by viewing the Milky Way from 10 million light years away. Then you slowly move through space towards the Earth in decreasing orders of magnitude until you reach a tall oak tree just outside the buildings of the National High Magnetic Field Laboratory in Tallahassee, Florida. After that, you progress from the actual size of a leaf into a microscopic world that reveals leaf cell walls, the cell nucleus, chromatin, DNA and, finally, into the subatomic universe of electrons and protons. Each picture is an image of something that is ten times smaller (or bigger) than the one preceding or following it. The number that appears on the lower right just below each image is the size of the object in the picture. On the lower left is the same number written as a power of ten. It is possible to buy the site as a screensaver.

The original 'powers of ten' concept was advanced by the architect Charles Eames, who first utilised powers to aid in visualisation of large numbers in 1952, and later, along with his wife Ray, directed a film having this title.

Other useful sites

The Internet provides access to a multitude of resources for science teaching. The following sites will provide you with an indication of the range of material available.

- **CHEMIX** provides downloadable chemistry education software for students and teachers (at secondary, advanced and university level). Included are CHEMIX calculator tools: molecular (handles isotopes, amounts), thermochemical, balance (stoichiometric), solubility product, weak acid/base, buffer, electrochemistry; also a periodic table (history, 22 physical properties, NMR data) and a dictionary.
- *Dataloggerama* has been produced by Roger Frost (co-editor of this book), known for his work on IT and science in books, journals and presentations. The website features data files that can be downloaded, suggestions for datalogging experiments, software reviews and advice about which datalogging equipment to consider purchasing. Extensive links are provided to manufacturers' sites and sources of CD-ROMs and software. This is an excellent reference and data site.
- **The Power of Photography** is a site recently launched by Pentax, which explores and explains the physics of light, particularly as it relates to photography. It includes some useful demonstrations, virtual experiments and explanations on reflection, refraction, lenses, dispersion and filters. Also included are teacher's notes.
- **History of Microbiology** is a site that includes information on the key developments and the key individuals. It could be used to allow students to develop an understanding of the microbiology timeline, leading to an appreciation of the

key events in the development of the subject. For example, was it ethical for Jenner to deliberately infect others with smallpox? The **Jenner Museum** site is another useful one, which has information about smallpox, the development of vaccination and immunology, links to subject specifications, and even a word search.

- **Radioactivity** is a site that could find uses with older students. Homework exercises, suitable links and some interactivity are included.
- **Safety Information Resources** includes downloadable line drawings of people and equipment on a wide range of safety topics, including electricity and heat. Students and teachers could customise posters and worksheets. These will grab attention and enhance displays in the school or college laboratory. This is a valuable resource for any science teacher.
- **Schoolscience** is a growing content site that includes a large number of free online resources showing the application of the science taught in secondary schools. The material consists of a number of e-sources – electronic resources that give information on up-to-date industrial applications. They include information, pictures, quick questions and some interactivity. The coverage is very wide, for example: Fossils into Fuels, The Chemistry of Steelmaking, History of Medicines, The Human Genome Project, Periodic Table, Proteins, The Science of Audio Systems, The Viagra Story. Some teachers' notes are available (in downloadable .pdf format). There is also the opportunity to submit feedback. This is a very useful site which shows the applications as seen from the side of industry. Site sponsors include many industrial and manufacturing companies, for example: ABPI, BAMA, British Energy, CDA, GSK, ICI, Institute of Petroleum, Institute of Physics, MRC, PPARC, SGM, Sony, Unilever.

7.4 How do we find what we want?

There are a large number of ways into the Internet. The examples discussed below each offer a different way of searching for the information you require. See pages 142–3 for addresses.

1. **Google**™ can be used to search a website of your choice. For example, 'access site: my.sheffcol.ac.uk' typed into Google will search the Sheffield College site for the word 'access'. This facility is excellent when you are confronted with a site that has an unwieldy structure and a poor search facility.

2. **Google**™ **Sets** allows you to enter a small part of a set – then the set is completed, links appear and away you go. For example, the inputting of 'cell' and 'nucleus' leads to over 50 related terms, each with associated links. This is a very useful facility, especially with Advanced level students.

3. **Physics.org** is a gateway from the Institute of Physics (IoP), which uses a powerful natural language programme to give an accurate and relevant answer by the use of a database of refereed resources that are guaranteed by the IoP. If you provide your age and knowledge of physics, the answers become even more focused.

4. **Search Engine Colossus** is an international directory of over 1000 search engines, from around 200 countries, many in the language(s) of the country in which they are located. It also groups search engines by theme.

5. The **Teacher Resource Exchange** is a growing resource which includes over 1000 science-based resources and ideas. These vary from straightforward lesson plans, to PowerPoint presentations and detailed simulations. The best of these resources are identified by a panel of moderators against agreed criteria and included in Curriculum Online.

6. **Weblinks** provides over 5000 educationally useful, annotated, categorised and searchable links; many are categorised as 'Science'. Winner of a year 2000 Becta/ Guardian Website Award, the site now receives over one million hits annually.

Finding images

There are many sites available that can provide useful images. **ClassroomClipart** contains hundreds of scientifically based images which can be used for educational purposes. Others may be found at **Google™ Image Search**, where more than 150 million are listed and indexed, although many are not copyright free (see Box 3).

Box 3. Referencing and copyright

If a student copies some text from a website and uses it in their written work, this text should be referenced. The usual manner is to copy the website address from the address box, paste it onto the document and add both the name of the site and the date accessed. For example:

Planet Science:UK Visited September 2004.
URL http://www.planet-science.com

The same should be done when a student uses an image. However, many images on the Web are examples of publishing without permission and therefore violate copyright law. The copyright held in digital images on the Web can be a complex set of 'layers' of ownership, depending on how old the image is, when its original copyright holder died, whether it has been published prior to its digital publishing, and whether the digital version is an adaptation (as opposed to a more direct 'copy') of an original photograph. My interpretation of the law is that the use of an image obtained from the Internet by school students is permitted because their work is produced for the purposes of examination (Section 32 of the *Copyright, Designs and Patents Act 1988*). Should you wish to use images for any other purpose (for example, on the school website), then copyright clearance should be sought.

Talking to our colleagues

The Internet contains a number of discussion forums suitable for teachers. These forums can provide an easy method of consulting your colleagues about teaching techniques, examination questions, laboratory design and other issues related to science teaching. Each discussion list allows access to an archive, hence it is possible to look back at areas of discussion from the past. Available discussion lists include the following.

- **Biotutor-L** is a list specifically for biology teachers in schools and colleges, run by the Department of Biological Sciences at the University of Central Lancashire. Subscription requires a headed letter to be sent to the List Organsier.
- **The Chemists' Net** is a discussion forum for chemistry teachers interested in using the Internet both as a resource and for communicating on matters relating to teaching chemistry.
- **Physics Teaching: News and Comment** (PTNC) is a UK-based discussion group. To join, send an email to 'PTNC-request@iop.org'. Leave the subject line blank and in the body of the message type: 'subscribe PTNC [your email address]'. You can then contribute, respond and receive.

◆ *Conclusion*

The use of the Internet can only grow as accessibility continues to increase. This ease of access will allow our students to become experts in the art of searching for and processing information, not simply repositories of facts and information. They will gain insight from many sources, not just their own school or college. Future developments in information terms are well illustrated by recent innovations from MIT, who are providing free, unlimited access to a vast range of teaching material.

This increase in students' information-gathering expertise can only increase their awareness of the importance of the world around them, of the importance of citizenship, and of the importance of a scientifically literate community. This gradual change should be welcomed – we will develop scientifically literate students who are adept at gathering and linking information before reaching considered decisions. The Internet, then, well used, can act as the catalyst for burgeoning a more scientifically literate society.

◆ *References*

A New View of Statistics
   www.sportsci.org/resource/stats
Copyright, Designs and Patents Act 1988
   www.hmso.gov.uk/acts/acts1988/Ukpga_19880048_en_
   4.htm

◆ *Web addresses*

Anglo-Australian Observatory
  www.aao.gov.au/images
Biotutor-L
  www.biology4all.com/biotutor.asp
Brightsparks
  www.planet-science.com/sciteach/bright_sparks
British Geological Society (BGS)
  www.bgs.ac.uk
CHEMIX software
  http://home.c2i.net/astandne
ClassroomClipart
  http://classroomclipart.com/cgi-bin/kids/
  imageFolio.cgi?direct=Science
Dataloggerama
  www.rogerfrost.com
Electricity Map
  www.eskimo.com/~billb/miscon/ele-map.html
Faulkes Telescope
  www.faulkes-telescope.com
Google™
  www.google.com
Google™ Image Search
  http://images.google.com
Google™ Sets
  http://labs.google.com/sets
Heart
  www.platolearning.co.uk/heart.asp
  Also available on the CD-ROM acccompanying this book.
History of Microbiology
  http://users.stlcc.edu/kkiser/History.page.html
Jenner Museum
  www.jennermuseum.com
Massachusetts Institute of Technology (MIT)
  http://web.mit.edu
Neighbourhood Statistics
  www.neighbourhood.statistics.gov.uk
Physics.org
  www.physics.org

Physics Teaching: News and Comment (PTNC)
  PTNC-request@iop.org
Powers of 10
  http://micro.magnet.fsu.edu/primer/java/scienceopticsu/
  powersof10
Radioactivity
  www.sheffcol.ac.uk/projects/access_online/Health_Science
  _Access/Chemistry_1/Radioactivity
Safety Information Resources
  http://siri.org/graphics
Schoolscience
  www.schoolscience.co.uk/content
Search Engine Colossus
  www.searchenginecolossus.com
See the International Space Station
  http://esa.heavens-above.com/esa/iss_step1.asp
SEED Science Centre
  www.seed.slb.com
Teacher Resource Exchange
  http://tre.ngfl.gov.uk
The Chemists' Net
  www.chemclub.com/teachers
The Environment in your Pocket 2003
  www.defra.gov.uk/environment/statistics/eiyp/intro.htm
The Essential Guide to Rocks
  www.bbc.co.uk/education/rocks
The Power of Photography
  www.pentaxeducation.co.uk
UK Pollution Data
  www.aeat.co.uk/netcen/airqual/bulletins/
  weeklygraphs.html#op1
Weblinks
  www.sheffcol.ac.uk/links

8 The Internet as a source of debate

Darren Forbes

◆ Introduction

The Internet is an excellent resource for finding out current information and ideas. It can provide students with all the facts and figures they need to be able to discuss important scientific issues in a number of ways. In this chapter we take a look at different ways of using information from the Internet to encourage debate, while keeping the focus on scientific learning. There are also some suggestions for keeping the students on the right track and avoiding the pitfalls of extended research.

Four ways are considered in which information can be used in discussing or debating scientific issues in the classroom:

◆ producing presentations
◆ group discussions
◆ role play
◆ writing for an audience.

We will take a look at each in turn, discussing the best approach to each task and then focusing on a specific activity to show how the approaches can be made to work.

8.1 Presentations

It is now common for students at secondary level to be asked to present their research information to the rest of the class, especially if they have been asked to produce a PowerPoint presentation. This approach has two advantages:

(i) it gives students recognition and instant feedback about their work, and
(ii) it builds their confidence in presenting to an audience.

There are also a few disadvantages:

(i) some students may be unable to use the software,
(ii) the production of presentations can take a lot of time, and
(iii) when viewing the shows there can be a large amount of repetition.

If students have difficulty with the software it may be possible to pair them with partners, but watch out for one student doing all of the work while the other just watches.

The repetitive aspect of presentations is certainly annoying; fifteen shows on the pros and cons of nuclear power can certainly tax the teacher, not to mention the students! A generally better alternative is to select three or four shows and have those presented. If you have been monitoring the work in progress you should be able to find four shows that are sufficiently different in content and style. You should, of course, attempt to choose different students for each activity. The class can then discuss the shows after they have been presented and agree on the scientific content while debating the opinions presented. Assessment of the other students' work depends on the network you use. If you can access the students' files you can view the show at your leisure or, alternatively, the students will have to provide you with the show on disc.

The time it takes to produce the shows can be reduced by providing the students with appropriate templates to build on. Having a range of templates available will also add to the variety of the reports produced. A guide to making templates can be found at the end of this chapter and some examples are included on the CD-ROM.

Presentation example: dihydrogen monoxide (DHMO)

The 'Dihydrogen Monoxide Research Division' website presents an opportunity for students to learn to recognise that factual information can be presented in a manipulative way. If you haven't viewed the site then stop reading now and have a look at it: www.dhmo.org.

The site echoes the style and language of numerous campaign sites and can catch out even some of the more able students. The 'facts' presented on the site are generally true but are manipulated to present an agenda to ban the supposedly lethal 'DHMO'.

There are a few ways to use this campaign to develop your students' skills at evaluating information.

One possibility is to set groups the task of producing a report on the dangers of DHMO. Don't let the students in on the trick. Ask them not to share ideas with other groups, so that even if one group catches on, the others may complete the task without doing so. Get one or two of the groups to present their reports; go for the most frightening and alarmist. Then present a report of your own, explaining what DHMO really is but pointing out that many of the 'facts' are true. This should then lead to a discussion about how scientific information can be presented with bias and the importance of judging the source of 'factual' information.

Another possibility is to get the students to discuss the way that 'evidence' is presented on the site. Give the students the truth from the beginning and tell them that their task is to explain how the site uses scientific facts to make the reader afraid of plain water. For example, some of the uses of DMHO are listed as:

- in nuclear power plants
- in many forms of cruel animal research
- as an additive in certain junk foods and other food products.

Ask the students to discuss why they think these examples have been chosen by the author and how they are designed to manipulate the reader. You could then ask them to work on a 'pro-DHMO' presentation to set aside people's fears. An example of a 'pro-DHMO' website can be found at www.armory.com/~crisper/DHMO.

8.2 Group discussions

Discussion can be a valuable way of developing a student's scientific and social skills. The abilities to hold rational arguments and to accept that other people may not share your opinions are very important. Whole class discussions tend to be a bit limited, except in the most able and well behaved classes. Maintaining order while one person speaks at a time and making sure that students are paying attention can become a full-time task and reduces the amount of debate. It is usually better to split the class up into smaller groups. The example below shows how a class can be split to carry out research and debate, and come back together after the task.

Discussion example: a mobile phone mast
You could set the students the task of debating whether or not to site a mobile phone antenna in the grounds of the school. The whole activity takes two one-hour lessons or, as an alternative, the research phase can be done as homework.

The class could be split into groups of eight or so; then two students in a group would be asked to represent the opinions of typical parents, another two the opinions of the local council or school governors, two more for the phone company, and the final two for the children. Having two students represent each point of view within the group is much better than just one. The students can support each other and don't become isolated in their opinions. The students also seem to take the debate less personally this way. Each pair could be provided with a different set of websites and other resources for finding evidence. If particular students have strong opinions you can allow them to work in the group that will express that opinion or, for a greater challenge to them, in a group with the opposite opinion!

The students can then collect information from the Internet. (See Chapter 7 for pointers to effective searching.) During this research phase you should monitor their progress and make sure that they are collecting useful scientific (and economic) information. When the information is collected each group gets together and debates the issue. Give them a limited time to come to a decision about the mast being placed. They need to write down their decision with valid reasons. While they do this you can wander between groups, ensuring that the debate is civil but active. I like to play the 'devil's advocate' and help out any representatives that seem to be having difficulty coming up with an argument. When all groups have made a decision you can hold a plenary session to find out whether the mast will be built after all.

8.3 Role play

It is sometimes useful to get the students to role-play debates in a fuller sense than the discussion style outlined above. The students not only try to represent a point of view but also try to represent the feelings and style of different characters. Allowing the students to 'get into character' can increase their enthusiasm, but they need to know their role clearly. As with the discussion-based activity, you may want to pair students together for extra support. Even with this precaution you will find a number of students find it difficult to play roles effectively. If this is the case you can still fall back on the simpler form of debate where the students just discuss the evidence.

It is very important to set the students' roles at the beginning of the research stage. Then you can ask them only to find information that supports their assigned point of view, while maintaining their ignorance of the views of others. This will save research time and also guarantee differing presentations, reports or opinions.

Role play example: a matter of life and death

This activity can really grip the students and can be very relevant. It works very well with more able students but can be simplified for less able groups. Often the students will find an anti-smoking or anti-drug lesson 'a bit of a lecture'. The role play can bring all of the hazards of drug abuse into focus in a much more active and imaginative way than a traditional 'find out the facts on smoking' task.

The scenario is this: a hospital has a large number of patients and insufficient funds to treat all of them. The students have to prioritise treatment of the patients on medical grounds alone. Some examples of patients awaiting treatment might be:

- a middle-aged woman injured in a car accident in need of multiple transplants
- a middle-aged male smoker in need of a heart transplant
- a much older (non-smoking) man competing for the heart transplant
- a young obese man in need of that very same heart
- a middle-aged alcoholic woman in need of a liver transplant
- a young smoker with kidney disease
- a heroin addict in need of a lung transplant
- a young girl in need of cornea implants
- an old woman in need of a hip replacement.

Obviously you can customise the list to include patients with needs that you have been studying. You might want to remove the addictions if you don't want the students to focus too much on the ethical issues. Students can be very judgemental about the lifestyles of others.

In this scenario you should probably avoid assigning the students the patients' roles, otherwise they will concentrate too much on drama and not enough on science. Instead they should play the doctors and hospital committee members.

The students are assigned the task of deciding who gets the treatment. Set a limit that only five patients can be treated due to cost or resource restraints. The students then have to find out about each of the problems, how it is treated and how effective the treatment is. They should then make informed judgements based only on medical criteria about who should be treated.

Split the class into groups of roughly ten students. Most of the students play doctors, presenting their individual patient cases to a committee which will decide who gets treated and who doesn't. Let three students in each group play the committee members. Assign each doctor a case to find out about while the committee members are assigned the task of making a list of general priorities for their hospital. The committee could look into the latest government initiatives to help them come up with their lists. After the research phase is completed the cases are presented one by one to the committee and then the committee makes and justifies its decisions. It is useful to emphasise to the doctors that a committee usually goes with the doctor who presents an argument based on good evidence, but a touch of passion and persuasion never goes amiss.

All of the groups can then get back together for a plenary in which the different treatment lists are compared and the whole class can discuss which list they think is the fairest.

8.4 Writing for an audience: leaflets and posters

Another way of assessing students' research and understanding is to get them to create a piece of work to explain what they have found out. This can range from a simple poster or leaflet to a large class display. To make this type of task more challenging the students should be asked to target the information to a specific audience; this could be linked to English or Literacy objectives.

The leaflets or posters can be assembled while the students are carrying out their research if you wish to save time. You will need to try to make sure that the students are absorbing the information and not just throwing it together.

Creating the leaflet or poster afterwards gives the students more time to consider and discuss the information, but this can be very time consuming. Whichever way you choose, you need to make it clear that the work will be assessed according to the scientific information it explains and not just how good it looks.

To make sure that this type of task runs smoothly, you really need to make sure that the students are comfortable with the software you want them to use. It is best to check with your ICT department to find out when the students are being introduced to different programs, so that you don't end up spending the whole lesson teaching the students how to use Word Art or how to make the text wrap around a picture. There are plenty of templates built into applications like Word or Publisher that will make the production easier, or you can produce your own.

An example writing task: renewable energy

This type of task is quite familiar but here the students have a sense of both teamwork and competition. It can be quite time consuming but it can be very rewarding – and you get a great display for open evening!

Give the students a map of a large island with a range of energy needs. This should be a really big map so that A4 posters can be placed around it.

Split the class into research teams with about six students in each team. Then get them to research the advantages and disadvantages of different energy resources to see if they would be suitable for the island. Provide a list of the types of resource you want them to look at, to give them a starting point.

Each research team then gets together and shares and discusses the information they have gathered. Each student within the team is then allocated the job of producing an A4 poster explaining the advantages of one type of energy resource and where it should be sited (you may have to help out with this).

Once these are complete, each research team can briefly present its energy scheme and give its advantages and disadvantages. The best scheme is put up for display or, if you have a lot of wall space, you can put up all of the alternative schemes.

8.5 Other possible debates

Students should be able to find a wide range of information on the Internet to inform the following debates. Primary sources of topical science information are the BBC News site and the *NewScientist* site (see page 156).

- **Are genetically modified foods safe and useful?** There are many arguments for and against the development for GM food. Students could produce a 'pros and cons' leaflet to inform consumers in a supermarket, or they could role-play a scene from the government's recent 'Big Debate' between consumers, farmers, scientists, government officials and environmentalists. They could look into organic food, too.
- **Have we evolved from apes?** Some students just don't get evolution. Some are under the impression that we have changed from 'monkeys' into humans quite recently. Students could present the evidence or collect information to re-enact debates of the past (hats and beards optional).
- **Are the continents moving?** Work on rocks is often seen as dull. It needn't be. Ask the students to present conclusive evidence that the continents are moving, with scientific explanation of how this can happen. This activity works really well if the students are allowed to make PowerPoint presentations. I've seen great shows including footage of volcanoes erupting, earthquakes, and the continents sliding around over millions of years.
- **What makes a diet healthy?** There's plenty of information available on what we should be eating. Younger students could look into the different food groups and explain what they are used for in the body; this is great for a huge class poster. Older students could look into dieting, and debate the health and safety issues of some modern diet programmes. They could also debate what foods should be available in the school canteen.
- **What's the point of space exploration?** The students can look into the costs and benefits of space exploration, manned or unmanned. This is a good way of getting Key Stage 4 students to revisit the topic of the solar system. The excellent pictures on the Internet should encourage them (see section 7.2).
- **Is there other life in the universe?** You could put the students into three teams, one trying to prove the existence of alien visitors, one debunking the idea, and one giving the scientific view, i.e. trying to estimate the chances.

8.6 Some practicalities

Keeping your resources up to date

Although the Internet has all the information you could ever ask for, it is dynamic and things have a habit of disappearing when least expected. In order for the activities to run smoothly you will have to maintain your resources.

A week or so before you try to run the activity, check that the websites are still where you expected them to be. If not, then do a quick search and you will probably find an equivalent website.

Update your worksheets and templates with new links. If you have produced all of your resources by computer then this shouldn't be too difficult. If you have an intranet site maintained by technicians then they should be able to do this job regularly for you.

Recording information during research

During their research, students will need to record in an appropriate way the information they find. The Internet contains so much information that the students may well become confused without some guidance from you.

Printing

Often the first response by a student on finding a suitable website will be to ask if they can print the page. Monitors in schools tend to be small and many students find it uncomfortable to read from them. Printing directly from a website, however, can be very wasteful. A single web page can span half a dozen A4 sheets, and a class in a print frenzy can empty a printer in minutes! Try to limit printing as much as possible, perhaps allowing only a single sheet per student.

Making notes

Some students, especially those lacking in IT confidence, will want to record information into their books. This is hugely time consuming unless they have an advanced strategy for noting the information. The best approach I have found is to record information in a 'mind map' or 'spider diagram'; these allow students to quickly note the key concepts and their relationships to each other.

Watch out for students copying reams of information into their books. They can be in such a rush to copy everything that they do not take in any of the information and seldom gain anything from the task.

Cutting and pasting

This approach is a better one to encourage as students gain confidence in using several programs at once. They can copy chosen text and images directly from the websites and transfer them into their documents (see Box 2 in Chapter 7, page 135). This can again, however, lead to students presenting vast amounts of information that they haven't read or understood. It is important to get the students to read, analyse and edit the information that they find. To this end they should be provided with some form of template to work from. These templates are discussed below.

Making and using templates

A research-based activity, like all others, should always have clear objectives. For a scientific debate the students need to collect scientific facts and opinions efficiently and present them to the teacher or group.

Templates guide the students to their goals; they provide them with the basic building blocks on which to build their reports or arguments. Templates also allow for differentiation between groups or students in the groups. One template can have much more guidance than another; one can ask for only a few facts while another can get the students to give their own opinions or to evaluate the sources of the information.

A good template, whether computer or paper based, will allow the students to collect information more quickly, more accurately and with more purpose. In short, a template can make the difference between a well structured, well organised and successful learning experience and an Internet 'free for all'.

PowerPoint templates

You should try to keep these as simple as possible. If you develop a standard template format, the students will get used to it through the years and become much more efficient at producing presentations. The template should contain an 'instructions' slide for the students to refer to as they are piecing together information. 'Callouts' are a good way of guiding students as to what kind of information should appear on each slide. (These are little text boxes with arrows that point to places on the slides, added from the Drawing menu; you can see some of them in the example templates. I usually use ones that are shaped like 'thought bubbles' but there are other shapes.) These don't interfere with the handling or general layout of the slide and are easy for the students to delete once they have completed the task.

Don't be tempted to put complicated formatting into the template for the students. This makes the slide harder for them to use and it removes some of the more creative elements.

Word templates

These should also be of a simple format but have a few building blocks to help the students. You could provide a sheet with headings as questions and spaces for students to add in the answers. The sheet can also contain a few web links to sites where students can find the answers. I like to place a few images in the files to start the students off, too; this is a great help to less able students.

Paper templates

Computerised templates are not the only ones that can help students. Providing a simple table for the students to record information saves them a lot of time. Differentiated sheets or sheets with different tasks for different groups can be produced. It is again very useful to put some search terms or web page addresses on the sheet to give the students a head start.

A blank or partially completed 'mind map' or 'spider diagram' can be provided if the students are able to use this technique. It can be a real time-saver once students understand the method.

If classroom space permits I would recommend producing these resources on A3 paper to give the students plenty of space to make notes, but computers are seldom spaced out far enough for this to work in practice.

Tips on making templates

Making templates is time consuming. Even with practice it can take a couple of hours to make a template for PowerPoint. There are three ways to save a lot of time:

- **Using published teaching schemes.** Most new textbook courses have an accompanying CD-ROM or website with ICT resources. Even if you haven't bought a set of textbooks it is worth having a look at these published resources to see if there are any templates that can be made use of. You may have to edit them a little to remove textbook page references but this is certainly much quicker than starting from scratch. The disadvantage is that you have to pay.

- **Finding material on the Internet.** There are thousands of resources online. These may be in resource centres like the Teacher Resource Exchange, or just lingering on other schools' websites as examples of teachers' or students' work. Some of these can provide a very useful staring point. You may find a report on river pollution, for example. With a little work you can strip out some, or most, of the content and produce a suitable template or two. Leaving in a few of the pictures gives the students an idea of what sort of images would look good. This kind of reproduction is generally acceptable as long as you don't publish the work as your own.

- **Sharing.** If you've made something really good, then share it! The four hours you've spent making a set of role-play templates can be your gift to the teaching community. Obviously you will share resources with the rest of your department, but also try to put your work somewhere accessible on the Internet. Hundreds of teachers will be grateful, although you may never know it!

◆ *Conclusion*

This chapter has outlined the benefits and pitfalls of using the Internet to inform scientific discussions in the classroom. These kinds of activities, although potentially time consuming, are amongst the most important ones that students can take part in. Out of the 30 or so students in each class, only a few will go on to be scientists or engineers, but *all* of them will need to be able to discuss their medical treatment with a doctor, or make up their mind about what they should be eating, where their electricity should come from, whether we should be developing *this* weapon or *that* genetic cure, and so on. With practice in the skills needed to make informed and rational decisions, these students stand a much better chance of becoming the thinking adults we as teachers would wish them to be.

◆ *Resources*

BBC News
   www.bbc.co.uk/news
DHMO website
   www.dhmo.org
NewScientist
   www.newscientist.com
Teacher Resource Exchange
   http://tre.ngfl.gov.uk

Index

INDEX